Fundamentals of Mass Determination

Michael Borys · Roman Schwartz
Arthur Reichmuth[†] · Roland Nater

Fundamentals of Mass Determination

Springer

Michael Borys
Fachlabor 1.41
Physikalisch-Technische Bundesanstalt
Bundesallee 100 38116
Braunschweig
Germany

Roman Schwartz
Physikalisch-Technische Bundesanstalt
Abteilung Akustik
Braunschweig
Germany

Arthur Reichmuth[†]
Mettler-Toledo AG
Im Langacher
8606 Greifensee
Switzerland

Roland Nater
Mettler-Toledo International Inc.
Im Langacher
8606 Greifensee
Switzerland

ISBN 978-3-642-43300-9   ISBN 978-3-642-11937-8 (eBook)
DOI 10.1007/978-3-642-11937-8
Springer Heidelberg New York Dordrecht London

Pre-Press: Werner Brunner, Mettler-Toledo International Inc., Global MarCom Greifensee, Switzerland
Cover Photo by Mettler-Toledo AG, Greifensee, Switzerland, Photographer: Markus Weber, Zürich, Switzerland

Printed on acid-free paper

Springer is part of Springer Science+Business Media (www.springer.com)

Preface

This booklet is based on the MT brochure "Fundamentals of Mass Determination" published in 1991. It has now been updated and completely revised on the basis of the PTB report "Guide to mass determination with high accuracy" published in 2006.

In this work, the definition and dissemination of the unit of mass is explained, starting with an introduction to metrology and mass determination. Establishing a mass scale requires corresponding mass standards and mass comparators. The metrological requirements for weighing instruments, weights, and measuring conditions are explained and discussed, based on international directives and the applicable legal metrology regulations. International directives and institutions are striving towards the world-wide uniform implementation of these requirements. Processes used to determine density and volume are described to the extent that they apply to mass determination. Calculating measurement uncertainty requires the consideration of the effect of influencing quantities on mass determination. An overview of this topic is provided to make it easier to determine and specify measurement uncertainty in practice. Additional information in the form of tables, illustrations, and literature references allows the reader to extend the study of mass metrology.

Dr. Michael Borys, PTB
Dr. Roman Schwartz, PTB
Arthur Reichmuth, MTG
Roland Nater, MTII
Braunschweig (DE) and Greifensee (CH)   March 2012

Foreword

Within the covers of Fundamentals of Mass Determination, the reader will find both a general introduction to and a basic reference for the field of mass metrology. The authors begin with the definition of the kilogram unit in the International System of Units (SI) and then proceed to describe the establishment of a mass scale based on a 1 kg standard. This process starts with suitable mass standards, mass comparators, and an understanding of how these components are used. The mass scale is then disseminated to a wide group of users in science, legal metrology, and industry. This is accomplished in large part by establishing metrological requirements for weighing instruments, weights, and measuring conditions. Important international directives and the requirements of legal metrology are discussed. The processes used to determine density and volume are described to the extent that they apply to mass determination. The essential topic of measurement uncertainty is introduced, including the influence of various experimental parameters. The authors succeed in presenting a surprising level of technical information in a short and readable book. Additional information is provided in the form of tables, illustrations and literature references so that the reader can pursue these subjects further.

Dr. R.S. Davis
Consultant BIPM   March 2012

Table of contents

1. Introduction

Determining or estimating mass is common in physics, chemistry, and astronomy, using various measurement procedures in a wide mass range, for example, the mass of an electron is approximately 10^{-30} kg, and the mass of the sun is approximately 10^{+30} kg.

The following relations of the gravitational potentials

$$\frac{m_{g1}}{m_{g2}} = \frac{F_1}{F_2} \cdot \frac{r_1}{r_2} \tag{1.1}$$

($m_{g1,2}$ active gravitational mass, $F_{1,2}$ gravitational potential, $r_{1,2}$ distance), weight forces

$$\frac{m_1}{m_2} = \frac{F_{G1}}{F_{G2}} \tag{1.2}$$

($m_{1,2}$ passive heavy mass, $F_{G1,2}$ weight force), or accelerations

$$\frac{m_{i1}}{m_{i2}} = \frac{a_1}{a_2} \tag{1.3}$$

($m_{i1,2}$ inert mass, $a_{1,2}$ acceleration), and non-mechanical procedures (e.g. according to the radiometric measurement principle) are applied here.
Within the scope of the realisation and dissemination of the unit of mass, questions of mass determination by means of force comparisons that can be carried out in the range from 10^{-6} kg to 10^{+4} kg, e.g. by weighing, are of particular interest.

1.1 Mass and metrology

Metrology is the science of measurement and its application [1]. It includes both theoretical and practical measurement considerations, regardless of the respective measurement uncertainty or the respective scientific or technical field of study. The terms metrology and measurement (technique) are often used synonymously.
Metrology is embedded in an infrastructure that includes measuring, standardisation, verification, and quality assurance. As a branch of general metrology, legal metrology includes all areas that cover legal requirements for measurements, units, measuring devices, and measurement procedures [2]. Ensuring the uniformity and validity of measurements and measuring devices in legal metrology

M. Borys et al., *Fundamentals of Mass Determination*,
DOI: 10.1007/978-3-642-11937-8_1, © Springer-Verlag Berlin Heidelberg 2012

requires scientific, technical, legal, administrative, and organisational prerequisites.

Mass is one of the seven base quantities of the International System of Units (Système International d'Unités, abbreviated as SI). The significance of mass and its determination within metrology, especially legal metrology, is derived from its direct influence on many areas of measurement as well as current quality management, consumer protection, and product liability requirements.

1.2 The International System of units (SI) and the SI base unit "kilogram"

1.2.1 The International System of units (SI)

The International System of Units – abbreviated as SI in all languages – was introduced by the 11th General Conference on Weights and Measures (CGPM) in 1960 [3]. The International System of Units has only one unit for each physical quantity. SI units are classified as base units and derived units. The general conference decided to build the International System of Units on seven selected base units (Table 1.1). In addition to the condition that the remaining units can be derived from the base units, the selection of seven base units is based on both historical and pragmatic (i.e. not necessarily scientific) reasons. Derived units are calculated from the base units through multiplication and division according to the relationships of the corresponding quantities. They are derived coherently, e.g. using only the multiplier 1. For example, the Newton is the derived SI unit for force, $1 \text{ N} = 1 \text{ m kg s}^{-2}$.

The seven base units are commonly listed in the sequence shown in Table 1.1. This approach is also used for exponents. Deviations from this sequence are permitted in order to emphasise the origin of the derived unit (e.g. kg m^{-3} as a unit of density).

Table 1.1:
The seven SI base units

Name	Symbol	Quantity
metre	m	length
kilogram	kg	mass
second	s	time
ampere	A	electrical current
kelvin	K	thermodynamic temperature
mole	mol	amount of substance[1]
candela	cd	intensity of light

[1] Amount with regard to the number of elementary entities (e.g. atoms or molecules) contained, not the mass.

1.2.2 The SI base unit "kilogram"

At the end of the 18th century, radical changes in political conditions relating to the French Revolution offered the historical opportunity to also change a wide variety of weights and measures, harmonising them on a universal basis. The new units were to be based on natural, mutually agreed constant quantities and, as such, they would not be of advantage to any one nation [4]. Thus, the first corresponding law dated March 26, 1791, ratified by the French National Assembly and sanctioned by Louis XVI, reads in part: "In view of the fact that introducing uniform units of weights and measures requires a natural, constant unit of measure to be established, and that the only way to extend this uniformity to other nations and to induce them to agree on a measuring system is to select a unit that contains nothing arbitrary and nothing that is in any way specific to the situation of any nation on earth, ... the National Assembly adopts the quantity of a quarter of the earth's meridian as the basis of the new measuring system." [4, 5] In 1793, the metre was introduced as a unit of length defined as one ten-millionth of a quarter of the earth's meridian, the grave was fixed as the mass of one cubic decimetre of water, and the decimal scale was established with Greek and Latin prefixes [4, 6] (Table 1.2). In 1795, the units were already related to platinum prototypes that still had to be produced according to the results of the meridian measurements [4]. The initial unit of mass was the gram as the mass of one cubic centimetre of water at 0 °C, which was later replaced by the kilogram as the mass of one cubic decimetre of water at 4 °C. The prototypes of the metre and the kilogram were sanctioned in 1799 and stored in the archives of the Republic of France. This kilogram prototype is still referred to as the "Kilogramme des Archives"[2] even today [4]. Since the first General Conference on Weights and Measures in 1889, the unit of mass has been defined as the mass of the international prototype of the kilogram stored at the Bureau International des Poids et Mesures (BIPM) in Sèvres near Paris (Figure 1.1). The international prototype of the kilogram and two other 1 kg cylinders were made of a platinum-iridium alloy (Pt-Ir) in 1878/79 according to the then latest melting and refining processes, and adjusted to the mass of the "Kilogramme des Archives" in 1880. The diameter and height of this cylinder, which consists of 90 % platinum and 10 % iridium, are 39 mm respectively.

Figure 1.1:
International prototype of the kilogram, stored at the BIPM in Sèvres near Paris (Image by courtesy of BIPM, Sèvres, F)

---

[2] French for "kilogram of the archive".

The Pt-Ir alloy has a density of 21.5 g/cm³. The national kilogram prototypes in the member states of the Metre Convention are copies of the international prototype of the kilogram. They consist of the same material and have the same shape and dimensions. According to the definition from 1889, the kilogram is now the only one of the seven base units that is not defined by an atomic or fundamental constant but by a physical embodiment (artefact). Therefore, it is not only exposed to risks (such as damage or loss), but is also subject to temporal changes. Throughout the period from 1988 to 1992, the national prototypes, the reference standards, and the working standards of the BIPM were compared to the international prototype for the third time. This comparison showed that the differences in mass between the international prototype and the national proto-types, the reference standards and the working standards of the BIPM, have increased by an average in the order of 50 µg over the last 100 years. Unfortunately there is currently no way to prove the drift of the international kilogram proto-type with sufficient accuracy through independent measure-ments. Since around the beginning of the 1970s, several sophisticated experiments have been established all over the world in order to relate the kilogram to an atomic or funda-mental constant with a high degree of accuracy (e.g. Avoga-dro project, watt balance, ion accumulation) [7]. However, the relative standard measurement uncertainty of 2×10^{-8} required to verify the stability of the international prototype of the kilogram in a reasonable period of time and to redefine the unit of mass has not been achieved to date. For more information about the current discussion of the redefinition of the kilogram it is referred to [8–11].

Table 1.2:
Decimal submultiples and multiples of the mass unit

Unit	Symbol	Relationship to the base unit
nanogram	ng	$1 \text{ ng} = 10^{-12} \text{ kg}$
microgram	µg	$1 \text{ µg} = 10^{-9} \text{ kg}$
milligram	mg	$1 \text{ mg} = 10^{-6} \text{ kg}$
gram	g	$1 \text{ g} = 10^{-3} \text{ kg}$
kilogram	kg	base unit
ton	t	$1 \text{ t} = 10^{3} \text{ kg}$

1.2.3 Units outside the SI system

Some Anglo-Saxon countries still use two other systems for units of mass: the Avoirdupois system and the Troy system. Both systems use the "grain" (gr) as the base:

1 grain = 64.79891 mg (exact)   (1.4)

For the Avoirdupois system:

1 pound = 16 ounces = 7000 grains
($\approx$ 453.59237 g) (1.5)

For the Troy system:

1 troy pound = 12 troy ounces = 5760 grains
($\approx$ 373.24172 g) (1.6)

These and all other units that do not conform to the SI system are prohibited as legal units in all member states of the European Community (except for Great Britain).

1.2.4 Units and names of units that are only permitted for special areas of application

The metric carat, abbreviated as "ct" [12], is also permitted to determine the mass of precious stones.

1 ct = 2×10^{-4} kg = 0.2 g (1.7)

The length dimensions of textile fibres and threads are specified in tex, abbreviated as "tex".

1 tex = 10^{-6} kg m^{-1} = 1 g/km (1.8)

1.2.5 Units defined independently of the seven SI base units

The atomic unit of mass (symbol "u") is one-twelfth of the mass of an atom of the nuclide 12C and is, therefore, defined independently of the seven SI base units. It can be used in conjunction with SI units and can be converted into the SI unit kilogram with the equation [13]

1 u = $1.660\ 538\ 782 \times 10^{-27}$ kg. (1.9)

1.3 Fundamentals and basic terms

Weighing instruments and weights have been used to determine the "quantity of matter" as "the measure of the same" [14] for around 5000 years [7]. Although various descriptions and information we now associate with the term "mass" were substantiated much earlier, the modern understanding of mass as a physical quantity was not formed until the development of Newton's mechanics. Measuring a physical quantity requires the determination of its quantity relative to a comparative quantity, its unit (in this case, the kilogram embodied by mass standards or weights) by means of a measuring device (in this case, the weighing instrument, namely a balance or scale) and a measurement procedure (in this case, the weighing process).

The basic mass determination terms are explained below (see [15]).

Mass

Mass m describes the attribute of a body which expresses itself in terms of inertia with regard to changes in its state of motion as well as its attraction to other bodies.

Weighing value

When weighing in a fluid (liquid or gas) with a density of ρ_{fl}, the weighing value m_w is defined by the equation

$$m_w = m \frac{1 - \rho_{fl}/\rho}{1 - \rho_{fl}/\rho_w} \ . \tag{1.10}$$

Here m is the mass of the weighed object, ρ is the density of the weighed object, and ρ_w is the density of the weights. The weighing value of the weighed object (or of a product) is equal to the mass of the weights that keep the instrument balanced or cause the same indication of the instrument.

Conventional mass

The conventional mass m_c is calculated using equation (1.10) with the standard conditions $\rho_{fl} = \rho_0 = 1.2 \ kg/m^3$ and $\rho_w = \rho_c = 8000 \ kg/m^3$. ρ is replaced by the density of the weighed object at 20 °C.

Force

Force F is the product of the mass m of a body and the acceleration a it experiences or would experience due to the force F:

$$F = m \, a \tag{1.11}$$

Weight force

The weight force F_G of a body with mass m is the product of mass m and gravitational acceleration g:

$$F_G = m \, g \tag{1.12}$$

Weight

The word "weight" is used primarily with three different meanings:

  a) as an alternative term to weighing value
  b) as a short form for weight force
  c) as a short form for weights

Whenever confusion is possible, the specific terms weighing value, weight force, or weights should be used in place of the word "weight".

Load

The word "load" has various meanings in technology (e.g. for power, force, or for an object). Whenever confusion is possible, the word "load" should be avoided because it has no specific meaning in metrology.

Figure 2.1:
Kilogram prototype number 52
(stored under two glass covers)

2. Mass determination

2.1 Dissemination of the unit of mass

In 1883, from three Pt-Ir prototypes designated as KI, KII, and KIII, the prototype KIII was chosen by the CIPM as the international prototype of the kilogram and designated with the Gothic letter $\mathbf{K}$. At the first General Conference on Weights and Measures in the year 1889, 30 of the first 42 available kilogram prototypes were distributed by draw among the member states and the BIPM. Two additional prototypes (KI and No. 1) were presented to the BIPM for safekeeping as reference standards (témoins) in conjunction with the IKP. The remaining prototypes were retained for subsequent allocation and stored at the BIPM. The BIPM received two additional prototypes as working standards, France received five prototypes, several states two prototypes each, and the remaining states one prototype each [4]. In the years 1929 to 1993, 40 more prototypes were produced; most of these were distributed to additional member states of the Metre Convention as national prototypes [16]. For example, Switzerland has the prototype number 38; the national mass standard acquired in 1954 by the Federal Republic of Germany is prototype number 52 (Figure 2.1).

According to a recalibration performed at the BIPM in 2010, the mass of kilogram prototype number 52 was determined to be $m_{52} = 1.000\ 000\ 292$ kg with a standard measurement uncertainty of $u(m_{52}) = 6 \times 10^{-9}$ kg.

Since the definition and realisation of the unit of mass is tied to a specific object, the unit of mass cannot be disseminated with higher accuracy than that permitted by mass comparisons with the international prototype. This explains the hierarchical structure of mass standards, which guarantees dissemination of the unit of mass with the highest possible level of accuracy.

2.1.1 Hierarchy of mass standards

The international prototype of the kilogram is at the top of the hierarchy for the dissemination of the unit of mass. The national prototypes are linked to the BIPM working standards, which are, in turn, linked to the international kilogram prototype (Figure 2.2). Therefore, the international kilogram prototype only had to be used in 1889, 1939, 1946, and most recently 1989/92 as a reference, and, thus, it is protected to a large extent against wear and possible damage. The unit is further disseminated by

M. Borys et al., *Fundamentals of Mass Determination*,
DOI: 10.1007/978-3-642-11937-8_2, © Springer-Verlag Berlin Heidelberg 2012

the respective national metrology institute, e.g. PTB, using stainless steel secondary standards (density 8000 kg/m³). From a metrological point of view, linking these secondary standards to the national prototype is the most difficult step, since the required transition from a density of 21500 kg/m³ (Pt-Ir) to 8000 kg/m³ (steel) causes a larger uncertainty of the air buoyancy correction than the uncertainty of weighing and other influencing quantities. The reference standards of verification authorities for legal metrology and the reference standards of calibration laboratories and other institutions are then linked to the PTB secondary standards. The prototypes, secondary standards and reference standards are standards of the highest accuracy. Handling always bears the risk of unexpected mass changes (e.g. due to wear, contamination) and possible damage. Therefore, the selected recalibration intervals should be as long as possible, but yet short enough so that significant changes in mass are recognised. The right-hand column of Figure 2.2 shows examples of the time between two recalibrations. Working and control standards are easier to replace. The interval at which they need to be verified or recalibrated depends on the conditions and frequency of use.

Figure 2.2:
Hierarchy of mass standards, using the Federal Republic of Germany as an example (Pt-Ir: platinum-iridium alloy, BIPM: Bureau Internationale des Poids et Mesures [International Bureau of Weights and Measures], CIPM: Comité International des Poids et Mesures [International Committee for Weights and Measures, Sèvres, F], PTB: Physikalisch-Technische Bundesanstalt)

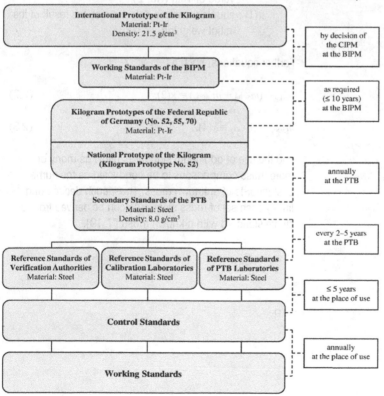

2.1.2 Mass scale

In order to determine the mass of arbitrary objects, multiples and submultiples of the mass unit must be realised in the form of mass standards and be linked to 1 kg standards. This is done by representing the nominal values in each decade, using a combination of standards. In legal metrology, the following nominal values shall be used for weight sequences in a set of weights: 1×10^n kg, 2×10^n kg and 5×10^n kg, $n \in \{..., -2, -1, 0, 1, 2, ...\}$ [17]. At least four standards are required per decade. The sequence 1, 2, 2, 5 is commonly used. In addition, the duplicate use of each nominal value, i.e. the use of six standards per decade with the values 1, 1, 2, 2, 5, 5 allows every nominal value in the mass scale to be covered twice [18]. Taking as an example the decade from 100 g to 1 kg, the first link-up weighing with a known mass m_{1kg} results in the equation

$$m_{1kg} - m'_{1kg} = x(1),\qquad (2.1)$$

where m_{1kg} is the mass of the standard with a nominal value of 1 kg (No. 1),
m'_{1kg} is the mass of the standard with a nominal value of 1 kg (No. 2),
$x(1)$ equals the mass difference as the result of the initial weighing.

Further equations, such as

$$m_{1kg} - (m_{500g} + m'_{500g}) = x(2),\qquad (2.2)$$

$$m_{500g} - m'_{500g} = x(4),\qquad (2.3)$$

and the use of additional standards allow as many or more mass comparisons to be conducted as the number of standards of unknown mass. Thus, each decade and finally each set of mass standards can be derived from a single standard with a known mass [7, 19].

Decade 100 g to 1 kg

Weighing	1 kg	1 kg	500 g	500 g	200 g	200 g	100 g	100 g
$x(1)$	+		−					
$x(2)$	+			−				
$x(3)$		+		−		−		
$x(4)$				+	−			
$x(5)$				+				
$x(6)$					+	−	−	−
$x(7)$					+	−		
$x(8)$					+			
$x(9)$						+	−	−
$x(10)$							+	−

Decade 10 g to 100 g

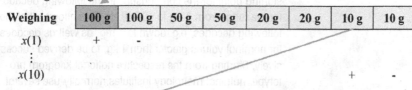

Weighing	100 g	100 g	50 g	50 g	20 g	20 g	10 g	10 g
$x(1)$	+	−						
$\vdots$								
$x(10)$							+	−

Decade 1 g to 10 g

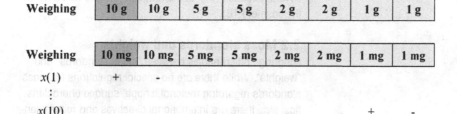

Weighing	10 g	10 g	5 g	5 g	2 g	2 g	1 g	1 g

Weighing	10 mg	10 mg	5 mg	5 mg	2 mg	2 mg	1 mg	1 mg
$x(1)$	+	−						
$\vdots$								
$x(10)$							+	−

and for larger nominal values

Decade 1 kg to 10 kg

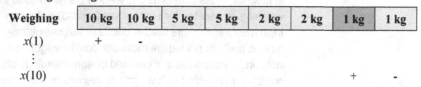

Weighing	10 kg	10 kg	5 kg	5 kg	2 kg	2 kg	1 kg	1 kg
$x(1)$	+	−						
$\vdots$								
$x(10)$							+	−

Figure 2.3:
Example of derivation of a mass scale from a weighing scheme with seven unknown standards and ten weighings per decade [18]

Various weighing schemes can be used depending on the requirements and the specified weight sequence in a set of weights. Figure 2.3 illustrates an example with seven unknown standards (with the weight sequence 1, 1, 2, 2, 5, 5, 10) and ten weighings per decade. The first line shows that during the initial weighing, the known 1 kg standard (symbol "+") is compared to the unknown 1 kg standard (symbol "–"). The weighing result of this comparison is $x(1)$. The over-determined equation system with ten equations and seven unknowns allows the unknown mass values of the individual standards to be calculated with the aid of a least-squares adjustment (Appendix A.2). One of the 100 g standards that were determined in the first decade is the starting point for the comparisons in the following decade for the range from 10 g to 100 g, etc. This allows all of the following decades, e.g. down to 1 mg, as well as decades for nominal values greater than 1 kg, to be derived successively. Starting from the respective national kilogram prototype, national metrology institutes normally use several sets of mass standards to derive the mass scale across several decades (e.g. in the range of 1 mg to 5000 kg in the case of PTB).

2.2 Mass standards and weights

Language differentiates between "mass standards" and "weights". While there are no special regulations for mass standards regarding material, shape, surface characteristics, etc., there are international directives and recommendations that apply to weights as well as national regulations that establish error limits, materials, shapes, etc. [17, 20–22].

In practice, mass standards and weights are rarely used for weighing. In general, the user utilises (verified) weighing instruments which (are used in official or business transactions and) do not require mass standards (weights) for weighing. Instruments that are used in legal metrology are adjusted and verified with weights at intervals of one to four years. Comprehensive procedures (type approval, regular verification of the mass standards and weights used for adjustment and verification by verification offices and national metrology institutes) are in place to ensure that a verified instrument "measures correctly" even without weights. Therefore, mass standards and weights are primarily used to adjust and check weighing instruments and for precision mass determinations with relative uncertainties of $< 10^{-5}$ (see Section 2.3).

2.2.1 Conventional mass and maximum permissible errors

The conventional mass m_c of a weighed object with the mass m and the density ρ at a reference temperature of 20 °C corresponds to the mass of a standard with a density $\rho_c = 8000$ kg/m³, which it balances in air with a reference density of $\rho_0 = 1.2$ kg/m³. Therefore, the conventional mass is a function of the mass m and the density ρ (see equation 1.10) [15, 23]

$$m_c = m \frac{1 - \rho_0/\rho}{1 - \rho_0/\rho_c} = m \frac{1 - (1.2 \, \text{kg m}^{-3}/\rho)}{0.99985} . \qquad (2.4)$$

The conventional mass was introduced in order to reduce mass comparisons to a simple weighing process. With the introduction of standard conditions for the density of the weighed object ($\rho_c = 8000$ kg/m³) and the density of the air ($\rho_0 = 1.2$ kg/m³), reference conditions were defined for the adjustment of the instrument. The different weight forces of weighed objects with the same mass but different densities become comparable. If the density ρ of the weighed object deviates from the conventional density ρ_c, an instrument indicates the conventional mass when weighing under standard conditions in air ($\rho_a = \rho_0$). This weighing value corresponds to the same force action exerted on the instrument by a comparative mass m_c with density ρ_c at an air density of $\rho_a = \rho_0$. Since the conventional mass corresponds to the value of a comparative mass, the unit of the conventional mass is the kg. The mass m of a weighed object can be calculated from the conventional mass m_c using equation (2.4). The relative deviation of the mass from the conventional mass is

$$\frac{m - m_c}{m} = 1 - \frac{1 - \rho_0/\rho}{1 - \rho_0/\rho_c} = \frac{\rho_0}{m}(V - V_c) , \qquad (2.5)$$

with the "conventional volume" $V_c = m_c/\rho_c$. The term $\rho_0(V - V_c)$ corresponds precisely to the mass of the air with the density ρ_0 contained in the volume difference $(V - V_c)$. For a weighed object with a density of $\rho > 1000$ kg/m³, the relative deviation of mass from the conventional mass is less than 0.1 % (see Figure 2.4).

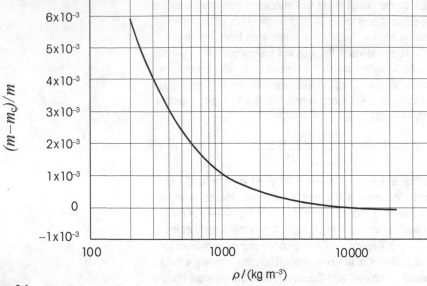

Figure 2.4:
Relative deviation of the mass from the conventional mass $(m-m_c)/m$ as a function of the density of the weighed object ρ (equation 2.5)

In legal metrology, weights are assigned to accuracy classes with defined error limits, otherwise called "maximum permissible errors" (mpe), according to international regulations (OIML R 111, Table 2.1). The specified nominal value of a weight is not the mass, but the conventional mass. The error limits also refer to the conventional mass. The accuracy class with the smallest mpe is class E_1. The mpe for subsequent classes with the designations E_2, F_1, F_2, M_1, M_2 and M_3 (with M_{1-2} and M_{2-3} as interim classes [17]) increase by a factor of approximately $\sqrt{10}$, respectively.

For each weight, the expanded measurement uncertainty U ($k = 2$) of the conventional mass must be less than or equal to one third of the specified margin of error δm.

$$U \leq \delta m / 3 \tag{2.6}$$

The expanded measurement uncertainty U is part of the mpe, i.e. the conventional mass m_c of a weight may not deviate from the nominal value m_0 by more than the difference between the specified margin of error δm and the expanded measurement uncertainty U.

$$m_0 - (\delta m - U) \leq m_c \leq m_0 + (\delta m - U) \tag{2.7}$$

In order to meet the uncertainty requirements when calibrating the conventional mass of weights of a given accuracy class, weights of a higher class (usually the next-higher) are used; for example, standards of class E_2 are used for weights of class F_1.

Nominal value	Maximum permissible errors δm in mg								
	Class E_1	Class E_2	Class F_1	Class F_2	Class M_1	Class M_{1-2}	Class M_2	Class M_{2-3}	Class M_3
5 000 kg			25 000	80 000	250 000	500 000	800 000	1 600 000	2 500 000
2 000 kg			10 000	30 000	100 000	200 000	300 000	600 000	1 000 000
1 000 kg		1 600	5 000	16 000	50 000	100 000	160 000	300 000	500 000
500 kg		800	2 500	8 000	25 000	50 000	80 000	160 000	250 000
200 kg		300	1 000	3 000	10 000	20 000	30 000	60 000	100 000
100 kg		160	500	1 600	5 000	10 000	16 000	30 000	50 000
50 kg	25	80	250	800	2 500	5 000	8 000	16 000	25 000
20 kg	10	30	100	300	1 000		3 000		10 000
10 kg	5.0	16	50	160	500		1 600		5 000
5 kg	2.5	8.0	25	80	250		800		2 500
2 kg	1.0	3.0	10	30	100		300		1 000
1 kg	0.5	1.6	5.0	16	50		160		500
500 g	0.25	0.8	2.5	8.0	25		80		250
200 g	0.10	0.3	1.0	3.0	10		30		100
100 g	0.05	0.16	0.5	1.6	5.0		16		50
50 g	0.03	0.10	0.3	1.0	3.0		10		30
20 g	0.025	0.08	0.25	0.8	2.5		8.0		25
10 g	0.020	0.06	0.20	0.6	2.0		6.0		20
5 g	0.016	0.05	0.16	0.5	1.6		5.0		16
2 g	0.012	0.04	0.12	0.4	1.2		4.0		12
1 g	0.010	0.03	0.10	0.3	1.0		3.0		10
500 mg	0.008	0.025	0.08	0.25	0.8		2.5		
200 mg	0.006	0.020	0.06	0.20	0.6		2.0		
100 mg	0.005	0.016	0.05	0.16	0.5		1.6		
50 mg	0.004	0.012	0.04	0.12	0.4				
20 mg	0.003	0.010	0.03	0.10	0.3				
10 mg	0.003	0.008	0.025	0.08	0.25				
5 mg	0.003	0.006	0.020	0.06	0.20				
2 mg	0.003	0.006	0.020	0.06	0.20				
1 mg	0.003	0.006	0.020	0.06	0.20				

Table 2.1: Maximum permissible errors ($\pm\delta m$ in mg) for the conventional mass of weights according to the international recommendation OIML R 111 [17]

2.2.2 Requirements

The requirements for weights refer to their physical and metrological characteristics. In order to ensure measurement trueness and stability that correspond to the respective accuracy requirements, the shape, dimensions, material, surface characteristics, density, magnetic properties, nominal values, and error limits of weights have been established in standards, directives, and ordinances [17, 20–22].

2.2.2.1 Shape

Weights must have a simple geometric shape without sharp edges or corners. In order to avoid deposits such as dust on the surface, they must not have any pronounced depressions. A high degree of stability, easy handling, and a favourable relationship between surface area and volume is ensured for mass standards and weights with a nominal value between 1 g and 20 kg by a cylindrical shape (Figure 2.5a) with a height-diameter ratio between $^3/_4$ to $^5/_4$. The block form with a fixed handle that does not protrude is also in widespread use for the range from 5 kg to 50 kg (Figure 2.5b). Weights with nominal values $\geq$ 50 kg are constructed so that, depending on the application, corresponding aids such as lifting and transportation equipment can be used safely and the weights can be stored securely.

Weights with nominal values < 1 g are shaped as polygonal plates or wires so they are easier to handle and to differentiate. The shape of weights that are not inscribed with their nominal value must correspond to Table 2.2.

Nominal value	Polygonal plates	Wires		
5, 50, 500 mg	Pentagon	Pentagon		5 segments
2, 20, 200 mg	Square	Square	or	2 segments
1, 10, 100 mg	Triangle	Triangle		1 segment

Table 2.2:
Shape of weights with nominal values
$\leq$ 1 g [17]

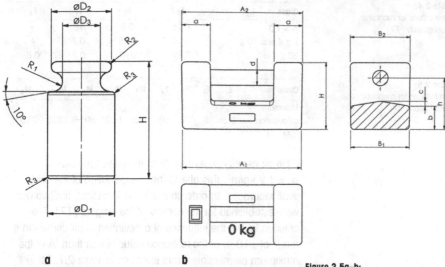

a   b

Figure 2.5a–b:
Examples of the design of weights
according to OIML R 111 [17].
a cylindrical weights,
b block-shaped weights

2.2.2.2 Material and surface properties

Weights must be made of a material that is highly resistant to corrosion caused by chemically and physically active substances in the atmosphere such as ozone, ammonia, oxygen, carbon dioxide, and water vapour. The material must have characteristics that ensure that changes in the mass of the weight that occur during normal use compared to the margin of error for the corresponding accuracy class can be disregarded. The surface of a weight must be smooth. Table 2.3 lists the maximum values for surface roughness according to the requirements specified in the OIML recommendation R 111. For weights with nominal values over 50 kg, twice the limit values in Table 2.3 apply.

Class	E_1	E_2	F_1	F_2
R_z / µm	0.5	1	2	5
R_a / µm	0.1	0.2	0.4	1

Table 2.3:
Limit values for surface roughness
[17]

The influence of the magnetic properties of a weight can be disregarded if the susceptibility and permanent magnetisation do not exceed the limit values specified in Table 2.4 and Table 2.5. The limit values were calculated so that, for commonly assumed maximum values of the magnetic flux density ($B_z = 110$ µT, $\partial B_z/\partial z = -34$ µT/cm, see [24]), the weighing results are not falsified by more than 10 % of the maximum permissible errors specified in Table 2.1.

Table 2.4:
Limit values for magnetic susceptibility [17]

Class	E₁	E₂	F₁	F₂
$m \leq 1\,g$	0.25	0.9	10	–
$2\,g \leq m \leq 10\,g$	0.06	0.18	0.7	4
$20\,g \leq m$	0.02	0.07	0.2	0.8

Table 2.5:
Limit values for magnetic polarisation [17]

Class	E₁	E₂	F₁	F₂	M₁	M₁₋₂	M₂	M₂₋₃	M₃
Maximum polarisation $\mu_0 M / \mu T$	2.5	8	25	80	250	500	800	1600	2500

If the air density ρ deviates from the reference value $\rho_0 = 1.2\ kg/m^3$, this affects the determination of the conventional mass. In order to minimise this effect, limit values were established for the density of the weights [17]. The criterion is that the influence of a deviation in air density in a range of $\pm 10\,\%$ of the reference value is less than $\frac{1}{4}$ of the maximum permissible errors specified in Table 2.1. Table 2.6 provides an overview of the resulting limit values for the individual accuracy classes.

In practice, the excellent material characteristics of austenitic steel with a density of 8000 kg/m³ has proven itself well (e.g. steel X1NiCrMoCu25-20-5, material number 1.4539).

Nominal value	$\rho_{min}, \rho_{max}\ (10^3\ kg\ m^{-3})$							
	Accuracy class (no specifications for class M₃)							
	E₁	E₂	F₁	F₂	M₁	M₁₋₂	M₂	M₂₋₃
≥100 g	7.934 – 8.067	7.81 – 8.21	7.39 – 8.73	6.4 – 10.7	≥4.4	≥3.0	≥2.3	≥1.5
50 g	7.92 – 8.08	7.74 – 8.28	7.27 – 8.89	6.0 – 12.0	≥4.0			
20 g	7.84 – 8.17	7.50 – 8.57	6.6 – 10.1	4.8 – 24.0	≥2.6			
10 g	7.74 – 8.28	7.27 – 8.89	6.0 – 12.0	≥4.0	≥2.0			
5 g	7.62 – 8.42	6.9 – 9.6	5.3 – 16.0	≥3.0				
2 g	7.27 – 8.89	6.0 – 12.0	≥4.0	≥2.0				
1 g	6.9 – 9.6	5.3 – 16.0	≥3.0					
500 mg	6.3 – 10.9	≥4.4	≥2.2					
200 mg	5.3 – 16.0	≥3.0						
100 mg	≥4.4							
50 mg	≥3.4							
20 mg	≥2.3							

Table 2.6: Lower and upper limit values for the density ρ_{min}, ρ_{max} [17]

2.2.2.3 Handling and cleaning

Maximum accuracy mass standards and weights must be treated with extreme care. The weights are stored in dust-proof boxes individually (usually from 1 kg and up) or as sets. They may only be handled with tweezers that have tips covered in plastic or another soft covering, weight forks made of wood, or with a clean, lint-free, non-greasing linen or leather cloth. Sets of weights are normally denominated so that each mass value can be represented by increments of the smallest weight in the set. According to OIML R 111 [17], the following increments are permitted:

$$(1, 1, 2, 5) \times 10^n \text{ kg}$$
$$(1, 1, 1, 2, 5) \times 10^n \text{ kg}$$
$$(1, 2, 2, 5) \times 10^n \text{ kg}$$
$$(1, 1, 2, 2, 5) \times 10^n \text{ kg}$$

The exponent n is a positive or negative whole number, or zero.

For weights and weight sets of the classes E_1 and E_2, a calibration certificate must always be issued for calibrations and tests performed, which are based on OIML R 111 [17]. For class E_2, such a certificate must include information about the conventional mass m_c, the expanded measurement uncertainty U, and the coverage factor k. In addition, certificates for weights of the class E_1 must include information about the density or volume of each weight as well as a statement indicating if these values were measured or estimated.

Weights must be handled and stored in such a way that they remain clean. Before using the weights, minor dust deposits must be removed with bellows or a soft brush. Cleaning must not remove material from the surface or deteriorate the surface characteristics of the weight (e.g. scratches). Other contamination – such as finger prints caused by improper handling – can be removed by cleaning all or part of the weight in pure alcohol, distilled water, or alternatively a non-detrimental solvent. Hollow weights must not be immersed in the solvent, so that liquid does not enter through the opening. Depending on the degree of contamination, cleaning can cause changes in mass that cannot be disregarded (e.g. through changes to the sorption layers). In order to determine and document the effect of cleaning, mass determination before and after cleaning is recommended. The stabilisation periods specified in Table 2.7 must be observed after cleaning with alcohol or distilled water. Since cleaning with alcohol has a greater effect on the sorption layers, the stabilisation periods are longer than those after cleaning with distilled water.

Class	E_1	E_2	F_1	F_2 to M_3
After cleaning with alcohol	7–10 days	3–6 days	1–2 days	1 hour
After cleaning with distilled water	4–6 days	2–3 days	1 day	1 hour

Table 2.7:
Stabilisation periods after cleaning [17]

Weights have to stabilise to the environmental conditions at the measurement location before calibration. The temperature difference compared to the weighing chamber should be as small as possible, especially for weights of the E and F classes. The required period depends on the temperature difference between the weight and the environment at the beginning of the stabilisation process as well as the size and the margin of error for the weight. Table 2.8 provides an overview of the minimum periods. Up to a nominal value of 5 kg, a stabilisation period of 24 hours is recommended as a practical guideline.

ΔT*	Nominal Value	Class E_1	Class E_2	Class F_1	Class F_2
	1000, 2000, 5000 kg	–	93**	79	7
	100, 200, 500 kg	–	70	33	4
	10, 20, 50 kg	45	27	12	3
±20 °C	1, 2, 5 kg	18	12	6	2
	100, 200, 500 g	8	5	3	1
	10, 20, 50 g	2	2	1	1
	< 10 g	1	1	1	0.5
	1000, 2000, 5000 kg	–	51**	1	1
	100, 200, 500 kg	–	40	2	1
	10, 20, 50 kg	36	18	4	1
±5 °C	1, 2, 5 kg	15	8	3	1
	100, 200, 500 g	6	4	2	0.5
	10, 20, 50 g	2	1	1	0.5
	< 10 g	0.5	0.5	0.5	0.5
	1000, 2000, 5000 kg	–	16**	1	0.5
	100, 200, 500 kg	–	16	1	0.5
	10, 20, 50 kg	27	10	1	0.5
±2 °C	1, 2, 5 kg	12	5	1	0.5
	100, 200, 500 g	5	3	1	0.5
	< 100 g	2	1	1	0.5
	1000, 2000, 5000 kg	–	–	–	–
	100, 200, 500 kg	–	1	0.5	0.5
	10, 20, 50 kg	11	1	0.5	0.5
±0.5 °C	1, 2, 5 kg	7	1	0.5	0.5
	100, 200, 500 g	3	1	0.5	0.5
	< 100 g	1	0.5	0.5	0.5

Table 2.8:
Minimum stabilisation periods in hours for temperature equalisation between the weight and the weighing chamber [17]

*ΔT = Temperature difference between the weight and the weighing chamber at the beginning of the stabilisation process
**Value not specified in OIML R 111 (2004), only valid for 1000 kg

2.3 Physical weighing principles and methods

Weighing normally compares weight forces. According to equation (1.12), the weight force F_G of a body is the product of its mass m and gravitational acceleration g. Weighing is based on this relationship. Therefore, the mass of two bodies is equal if they exert the same weight force at the same gravitational acceleration (e.g. at the same location). Weight forces can only be compared directly in a vacuum. In air, the weight and buoyancy forces are overlaid vectorially. For a balance with equal arms in equilibrium (Figure 2.6), this overlay leads to the torque equation

$$l_L (m_1 g_L - V_1 \rho_{aL} g_L) = l_R (m_2 g_R - V_2 \rho_{aR} g_R) \qquad (2.8)$$

with the designations

m_1, m_2 mass of the bodies (1 and 2),
V_1, V_2 volume (capacities) of the bodies (1 and 2),
g_L, g_R local gravitational acceleration (left and right),
l_L, l_R length of the effective lever arms (left and right),
ρ_1, ρ_2 density of the bodies (1 and 2),
ρ_{aL}, ρ_{aR} air density during weighing (left and right).

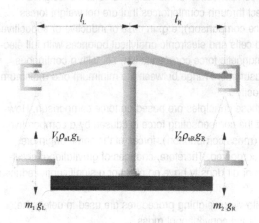

Figure 2.6: Forces and torques on a balance with equal arms

Spatial variations of the gravitational acceleration in the vicinity of the balance can normally be disregarded. With $g_L = g_R$, $l_L = l_R$, $\rho_{aL} = \rho_{aR} = \rho_a$ and $V_{1,2} = m_{1,2}/\rho_{1,2}$, it follows that the mass m_2 is

$$m_2 = m_1 \frac{1 - \rho_a / \rho_1}{1 - \rho_a / \rho_2} . \qquad (2.9)$$

In case of differences between the densities of the masses involved, the respective ratio between the air density and the solid density at the time of the comparison has a direct effect on the result of the mass determination. At a relative measurement uncertainty of up to 10^{-3}, a buoyancy correction is generally not required; therefore, the weighing value read from the instrument can be considered as the direct mass determination result (see also Sections 2.2.1, 2.5.1 and Figure 2.4).

The following physical principles are used in order to determine mass or the conventional mass using weighing instruments:

- Full compensation of the weight force of the weighed object by applying weights or mass standards (mass comparison), e.g. using a mechanical balance with equal arms, mechanical balances with built-in weights and mechanical substitution beam balances
- Partial compensation of the weight force of the weighed object with dial weights or permanently installed counterweights and additional fine compensation, e.g. with electromagnetic (electrodynamic) force compensation in case of electromechanical dial weight balances, or with built-in counterweights and partial compensation of the weight force through electromagnetic force compensation in case of electronic comparator balances
- Full compensation of the weight force of the weighed object through counterforces that are not weight forces (force comparison), e.g. in case of inductive or capacitive load cells and electronic analytical balances with full electromagnetic force compensation, i.e. with a continuous measurement range between the minimum and maximum capacity

All of these principles are based on force comparison. However, if the compensating force is caused by a comparative mass (mass comparison), forces of the same origin are being compared. Therefore, changes of gravitational acceleration or air density have no effect or a significantly reduced effect.

The following weighing procedures are used to determine mass or the conventional mass:

- In proportional or simple weighing, the weighed object is applied to the load receptor (load pan) after the instrument is zeroed and the mass (the conventional mass) is read.
- Differential weighing, i.e. the mass comparison of the weighed object with a mass standard (reference standard), using the transposition method (Gaussian weighing) is only possible on balances with equal arms. Here the specimen (the weighed object) and the reference standard are exchanged on the weighing pans at least once, and the results of both weighings are averaged.

- Differential weighing using the substitution method (Borda weighing) is possible on all types of instruments. Here the specimen (the weighed object) and the reference standard are compared on the same weighing pan in succession. With beam balances, the second weighing pan is loaded with a fixed auxiliary load (tare load).

For precision mass determinations with a relative uncertainty of $< 10^{-5}$, differential weighing is essential. Meanwhile, differential weighing using the substitution method has become the method of choice for modern comparator balances. It allows for simpler instrument designs and handling, resulting in shorter measurement times compared to the Gaussian method. In addition, the ability to automate the weighing process by using suitable weight-exchange mechanisms further increases measurement accuracy.

2.4 Scales and mass comparators

The first weighing instruments were simple balances. After the invention of the sliding weight scale, which is ascribed to the Romans, developments in the 19th century included mechanical weighbridges, crane scales, deflection scales, spring scales, and automatic weighing instruments. After the continued development of these mechanical weighing instruments up to the Second World War, the development of electromechanical instruments commenced; this led to a new variety of weighing instruments with various load cell principles [7].

2.4.1 Weighing instrument classifications

Analytical balances (laboratory balances) and comparator balances are used for high-accuracy mass determination. Analytical balances are instruments with a high resolution, where the scale interval d is usually less than or equal to the maximum capacity Max times 10^{-5}. The maximum capacity is usually no more than 10 kg. Verifiable analytical balances are classified as weighing instruments of special accuracy (OIML accuracy class I) and weighing instruments of high accuracy (OIML accuracy class II) [12, 25]. Analytical balances are frequently divided into the classes of instruments listed in Table 2.9, depending on the scale interval and maximum capacity [7, 26–28].

The term "comparator balance" or "mass comparator" has become commonly accepted for instruments with an even higher resolution, e.g. with a number of scale intervals of $n > 5 \times 10^7$ [26–28].

Designation	Common Maximum Capacity Max	Common Scale Interval d	Common Number of Scale Intervals $n = Max/d$
Precision balances	100 g ... 10 kg	1 mg ... 100 mg	10^4 ... 10^5
Weighing instruments of special accuracy			
• Macro balance	100 g ... 1 kg	100 µg	10^6 ... 10^7
• Semi-micro balance	25 g ... 100 g	10 µg	2.5×10^6 ... 10^7
• Micro balance	5 g ... 25 g	1 µg	5×10^6 ... 2.5×10^7
• Ultra-micro balance	≤ 5 g	0.1 µg	≤ 5×10^7

Table 2.9:
Common classification of analytical balances for high-accuracy mass determination

2.4.2 Mechanical balances with equal arms

Up to the beginning of the 20th century, the mechanical balance with equal arms represented the preferred design for analytical and laboratory balances; it is still used by some national metrology institutes today because of its high degree of measurement accuracy. Meanwhile, it is of next to no practical importance due to the disadvantages associated with its use (time-consuming measurement, complex operation, low comfort, sensitivity to vibrations and tilting) and the great amount of progress made with comparator balances with electromagnetic force compensation; therefore, we refer to additional literature here [7].

2.4.3 Electromechanical dial weight balances

Modern, high-resolution weighing instruments are equipped with an electromagnetic force compensation. The highest resolutions of up to 10^{-10} times the maximum capacity are achieved by electromechanical dial weight balances and electronic comparator balances with partial electromagnetic force compensation (the electrical weighing range is usually $10^5\ d$ to $10^6\ d$). Although electromechanical dial weight balances have been losing importance to electronic comparator balances since around 1990, they are still relatively common due to their ruggedness and very high accuracy in conjunction with a full weighing range. Figure 2.7 shows the basic structure of an electromagnetic dial weight balance.

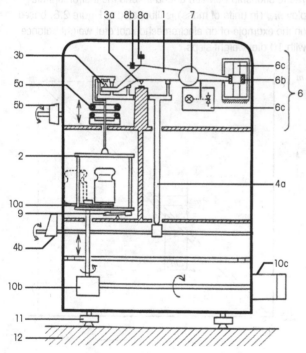

Figure 2.7:
Schematic illustration of a dial weight balance with electromagnetic force compensation and an automatic weight-exchange mechanism.
1 balance beam;
2 suspension;
3a main knife-edge,
3b secondary knife-edge;
4a, 4b, locking system, locking lever;
5a dial weights (including adjustment weight),
5b rotary switch for dial weights;
6 compensation system (electromagnetic force compensation) with:
6a electro-optical position sensor,
6b coil,
6c permanent magnet;
7 counterweight;
8a sensitivity adjustment,
8b zero point adjustment;
9 pan brake;
10a automatic turntable for weight-exchange mechanism,
10b lifting and rotating mechanism,
10c gear motor;
11 levelling screws;
12 weighing table.

The essential mechanical components are: The balance beam (1), the suspension (2), the main and secondary knife-edges (3a, 3b), the locking system and locking lever (4a, 4b), the dial weights (including the adjustment weight) with the rotary switches (5a, 5b), the fixed counterweight (7), the sensitivity and zero point adjustment (8a, 8b), and the pan brake to dampen the suspension oscillations after loading (9). The illustration also shows an automatic weight-exchange mechanism which consists of a turntable to hold several weights (10a), a lifting and rotating mechanism (10b) and a motor (10c).

The essential components of the electromagnetic compensation system are: The electro-optical position sensor (6a) consisting of the light source (LED), the light gap and the differential photodiode; the coil (6b) and the permanent magnet in a pot-type system (6c). The position sensor acts as a displacement transducer and the coil in the magnet system serves as an actuator for a PID controller that helps keep the vertical position of the balance arm at rest. Weight force differences are measured as proportional current changes. The electrical weighing range has to be adjusted with one or more adjustment weights – usually installed in the instruments – so that the weight force differences are indicated in units of mass.

The relationship between a load m' and the instrument display m_W (in units of mass) is illustrated in Figure 2.8, based on the example of an electromechanical dial weight balance with 10 dial weight steps.

Figure 2.8:
Adjustment characteristic of an electromechanical dial weight balance. 0–10 steps of the weight-dialing mechanism, m' load on the weighing pan (in units of mass), m'_{min} minimum capacity, m'_{max} maximum capacity, m_W balance indication (in units of mass), $m_{W,max}$ balance indication upper limit, δm_W linearity error, $\Delta m'_j$ electrical weighing range for the selected step j (here $j = 4$).

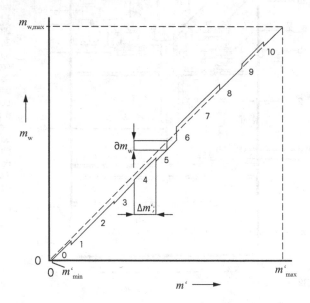

In this case the usable weighing range includes all loads between the minimum capacity (m'_{min}) and the maximum capacity (m'_{max}); the load is compensated by the dial weights combined with electromagnetic force compensation.

The largest positive or negative deviation from the theoretical linear curve shape is referred to as the linearity error δm_W. In case of dial weight balances, it mainly depends on the adjustment of the dial weights as illustrated in Figure 2.8. Therefore, substitution weighings are performed without changing the dial weight step and with an approximately constant load on the instrument in order to prevent linearity errors during high-accuracy mass determination (see Section 2.3).

2.4.4 Electronic analytical and comparator balances

Due to their high level of operator comfort, combined with a high resolution of up to 5×10^7 scale intervals, electronic analytical and comparator balances have prevailed over all other types of instruments where laboratory weighing technology is concerned. The high resolution can only be achieved with electromagnetic force compensation; it usually fully compensates for the weight force (entire measurement range from the minimum capacity to the maximum capacity). Built-in counterweights are also used in case of extremely high-accuracy requirements; in this case, the weight force is only partially compensated (restricted weighing range near the maximum capacity). Figure 2.9 shows the schematic structure of a top-loading electronic comparator balance with a fixed counterweight. The load receptor with the weighing pan (1) is guided by two parallel guide pairs so that only vertical movements are possible. The weight force is transmitted to the lever (6) via a gimbal-mounted load receptor (2), the pan carrier (3), a coupling element (4), and a flexible bearing (usually a cross flex bearing) (5).

A fixed counterweight (7) is located on the longer lever arm to mechanically compensate most of the effective weight force; some comparator balances also have several counterweights with different nominal values ("dial weights") which can be activated from the outside. The remaining weight force is electromagnetically compensated by a coil (9) that resides in the air gap of a permanent magnet system (8). As described above, the inductor current is controlled by an electro-optical position sensor (10–12) in relation to the load, so that the slit aperture (11) at the end of the longer lever arm remains in a defined rest position.

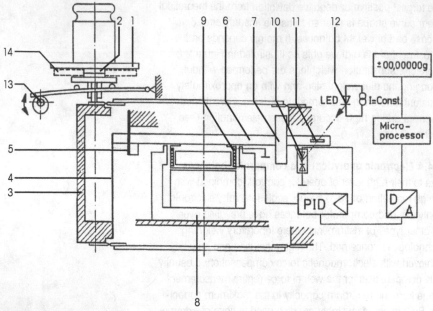

Figure 2.9:
Schematic illustration of a comparator
balance with fixed counterweight and
electromagnetic force comparison of
part of the weight force.
1 weighing pan,
2 gimbal-mounted load receptor,
3 pan carrier,
4 coupling element,
5 flexible bearing,
6 lever,
7 counterweight (fixed),
8 permanent magnet system,
9 compensation coil,
10 photodiode,
11 slit aperture,
12 light source (LED),
13 pan lock,
14 tare disk e.g. *Max*/2
(can be installed in addition).

Modern electromagnetic force compensation load cells have
a monolithic design (Figure 2.10). This manufacturing technique makes it possible to reduce the number of functional
components. The number of fine mechanical assembly and
adjustment steps is reduced and the reliability of the system
is improved.

Commercially available comparator balances (also see
Appendix A.3) can meet practically all metrology requirements. National metrology institutes and calibration laboratories can meet the requirements of the highest accuracy
class E1 according to the international OIML recommendation R 111. Using the example of PTB, Figure 2.11 illustrates
the uncertainty levels achieved for absolute mass determinations compared to error limits of the OML accuracy
classes [17].

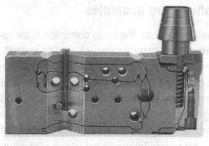

a

b

Figure 2.10:
Monolithic electromagnetic force compensation load cells produced with different manufacturing technologies.

a Monolithic load cell with mechanical elements produced by electric discharge machining (for greater visibility the model has been cut in the left one-third and the outer right part, Mettler Toledo);
b Monolithic load cell manufactured as a milled block system (Sartorius).

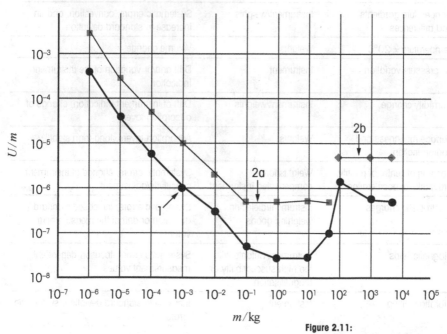

Figure 2.11:
Expanded measurement uncertainties ($k = 2$) of PTB for mass determinations according to the BIPM CMC tables[3] (1) and the error limits for classes E_1 (2a) and F_1 (2b) according to OIML R 111 [17]

---

[3] Calibration and measurement capabilities (CMC) [29]

2.5 Influencing quantities

During weighing in air, the main correction is the air buoyancy correction. However, other influencing quantities and disturbance factors must be considered in order to achieve a relative uncertainty $<1\times10^{-5}$ (Table 2.10). These quantities, which are usually linked to the environment, can affect the weighing instrument, the mass standards and the weighed object, respectively. The main influencing quantities and disturbance factors will be examined below, and measures to prevent or correct their undesirable effects will be discussed.

Influencing quantity/ disturbance factor	Effect on	Effect
Air density	Instrument/weights	Air buoyancy on instrument components and weights (systematic errors)
Temperature variation over time	Instrument	Drift of instrument indication, sensitivity
Temperature gradients and differences	Instrument/weights	Systematic errors, convection, and an increase in standard deviation
Temperature $\neq 20°C$	Weights	Volume change
Air pressure variation	Instrument	Drift and/or variation of the instrument indication
Humidity change	Instrument/weights	Drift of instrument indication, change of adsorption layers
Surface roughness and contamination	Weights	Adsorption layers, long-term stability
Position of centre of gravity, gravitational acceleration	Weights/force-compensated instrument	Systematic errors, change of instrument sensitivity/adjustment
Electrostatic charges	Instrument/dielectric weighing goods	Systematic errors, increased standard deviation or drift of the measurement values
Magnetic fields	Instrument/weights with too high susceptibility or magnetisation	Systematic errors, location-dependent measurement values
Vibration, tilting	Instrument	Increase in standard deviation, systematic errors
Eccentric load on the weighing pan	Instrument	Systematic errors

Table 2.10:
Influencing quantities and disturbance factors in high-accuracy mass determination

2.5.1 Air buoyancy correction

A mass comparison using substitution weighing in air for a standard with mass m_R (density ρ_R) and the specimen with mass m_T (density ρ_T) results in the weighing values

$$m_{WR} = m_R \frac{1-\rho_a/\rho_R}{1-\rho_0/\rho_c}\frac{1-\rho_0/\rho_J}{1-\rho_{aJ}/\rho_J} \, , \qquad (2.10)$$

$$m_{WT} = m_T \frac{1-\rho_a/\rho_T}{1-\rho_0/\rho_c}\frac{1-\rho_0/\rho_J}{1-\rho_{aJ}/\rho_J} \, . \qquad (2.11)$$

Here ρ_{aJ} is the air density during adjustment of the weighing instrument and ρ_J is the density of the weight used to adjust the instrument. Equations (2.10) and (2.11) result in the weighing equation for mass determination in air

$$m_T\left(1-\rho_a/\rho_T\right)= m_R\left(1-\rho_a/\rho_R\right)+\Delta m_W' \, , \qquad (2.12)$$

with

$$\Delta m_W' = \Delta m_W\left(1-\rho_0/\rho_c\right)\frac{1-\rho_{aJ}/\rho_J}{1-\rho_0/\rho_J} \, . \qquad (2.13)$$

$\Delta m'_W$ refers to the corrected and Δm_W to the uncorrected weighing difference, and ρ_a is the air density during the weighing. In the prevalent situation where the air density during adjustment is approximately $\rho_{aJ} = \rho_0$, the following applies:

$$\Delta m_W' = \Delta m_W\left(1-\rho_0/\rho_c\right)=0.99985\,\Delta m_W \, . \qquad (2.14)$$

This means that

$$\Delta m_W' = \Delta m_W \qquad (2.15)$$

can be substituted as an approximation for sufficiently small weighing differences.

If the densities ρ_T and ρ_R are replaced with the volumes V_T and V_R, then the weighing equation that corresponds to equation (2.12) is

$$m_T = m_R + \rho_a\left(V_T-V_R\right)+\Delta m_W' \, . \qquad (2.16)$$

The term $\rho_a(V_T-V_R)$ denotes the air buoyancy correction. Depending on the accuracy requirements, calculating the air buoyancy correction requires a more or less accurate determination of air density (see Appendix A.1) and possibly of the volume of the standards (see Section 3).

If auxiliary masses are used with the standard (m_{ZR}, V_{ZR}) and/or the specimen (m_{ZT}, V_{ZT}), the complete weighing equation is as follows:

$$m_T = m_R + m_{ZR} - m_{ZT} + \rho_a \left(V_T - V_R + V_{ZT} - V_{ZR}\right) + \Delta m'_W .$$
(2.17)

If the conventional mass m_{cT} of the specimen is being determined instead of the mass m_T (on an electromagnetically compensated analytical or comparator balance), the weighing equation

$$m_{cT} = m_{cR} \left(1 + C\right) + \Delta m''_W$$
(2.18)

can be derived with the instrument indication

$$m_{WR} = m_{cR} \frac{1 - \rho_a / \rho_R}{1 - \rho_0 / \rho} \frac{1 - \rho_0 / \rho_J}{1 - \rho_{aJ} / \rho_J}$$
(2.19)

and

$$m_{WT} = m_{cT} \frac{1 - \rho_a / \rho_T}{1 - \rho_0 / \rho} \frac{1 - \rho_0 / \rho_J}{1 - \rho_{aJ} / \rho_J} .$$
(2.20)

Here

$$C = \frac{(\rho_R - \rho_T)(\rho_a - \rho_0)}{(\rho_R - \rho_0)(\rho_T - \rho_a)}$$
(2.21)

is the air buoyancy correction, and

$$\Delta m''_W = \Delta m_W \frac{1 - \rho_{aJ} / \rho_J}{1 - \rho_0 / \rho_J} \frac{1 - \rho_0 / \rho_T}{1 - \rho_a / \rho_T}$$
(2.22)

is the corrected weighing difference. In many cases, the air buoyancy correction C is so small compared to the uncertainty that needs to be achieved for the specimen that it can be disregarded. Under normal environmental conditions, the relative correction of the weighing difference Δm_W according to equation (2.22) is usually also very small (generally $< 1.5 \times 10^{-5}$), so that it is not required if the condition

$$\Delta m_W < 30000 \, d$$
(2.23)

is met for the weighing difference. For $\rho_a = \rho_{aJ} = \rho_0$, the relation

$$m_{cT} = m_{cR} + \Delta m_W$$
(2.24)

applies exactly. This once again illustrates the advantage of the conventional mass. If the instrument adjustment and the determination of the conventional mass are performed at an air density of $\rho_0 = 1.2$ kg m^{-3}, air buoyancy correction – in contrast to mass determination – is not required!

2.5.2 Thermal influences

If the temperature in the measurement room is not constant over time, the temperature changes cause dimensional changes to the mechanical weighing system as well as changes to the electrical and magnetic characteristics of an electromagnetic force compensation system. Therefore, the result is not only a drift or variation of the instrument indication but also a change in the sensitivity of the instrument. Temperature changes or variations over time can be reduced, e.g. by suitable air conditioning of the measurement room or by using a basement room located below ground. Certain limit values for temperature changes in the laboratory can be derived, depending on the required measurement uncertainty (see Table 2.11). Temperature differences between the standard, the weight to be calibrated, and the instruments are especially critical, even for mass determinations with relative uncertainties of approximately 1×10^{-5}. If there is no thermal equilibrium, temperature gradients cause convection effects in the weighing chamber, which cannot only lead to an increased standard deviation but also especially to systematic errors in the weighing difference [30].

A sufficiently long waiting period (usually several hours) to allow the temperature of the specimen and the standard to adapt to the temperature in the weighing chamber can solve this problem (large weights can be left in the vicinity of the instrument under a common bell jar). In order to minimise temperature gradients in the weighing chamber, the exposure of weighing instruments and weights to direct thermal radiation (e.g. from solar radiation or radiators) should be avoided.

Specimen (calibration standard) T:					
U_T/m (100 g ≤ m ≤ 50 kg)	1.5×10^{-7}	5×10^{-7}	1.5×10^{-6}	5×10^{-6}	1.5×10^{-5}
Class	E_1	E_2	F_1	F_2	M_1
R_a / µm	0.1	0.2	0.4	1.0	–
R_z / µm	0.5	1.0	2.0	5.0	–
χ	0.02	0.07	0.2	0.8	–
$\mu_0 M_z$ / μT	2.5	8.0	25	80	250
Reference standard R:					
U_T/m (100 g ≤ m ≤ 50 kg)	$\leq 5 \times 10^{-8}$	$\leq 1.5 \times 10^{-7}$	$\leq 5 \times 10^{-7}$	$\leq 1.5 \times 10^{-6}$	$\leq 5 \times 10^{-6}$
Class	"E_0"[4]	E_1	E_2	F_1	F_2
R_a / µm	< 0.1	0.1	0.2	0.4	1.0
R_z / µm	< 0.5	0.5	1.0	2.0	5.0
χ	< 0.02	0.02	0.07	0.2	0.8
$\mu_0 M_z$ / μT	< 2.5	2.5	8.0	25	80
Volume V:					
U_{VT}/V_T ($m \geq$ 100 g)	3×10^{-4}	1×10^{-3}	3×10^{-3}	1×10^{-2}	3×10^{-2}
Measurement room:					
U_p / mbar	0.3	1.0	3.0	10	30
U_t / °C	0.1	0.2	0.5	0.5	1.0
U_{hr} / %	2	2	5	10	10
dt_{1h} / °C	± 0.3	± 0.7	± 1.5	± 2.0	± 3.0
dt_{12h} / °C	± 0.5	± 1.0	± 2.0	± 3.5	± 5.0
dh_{4h} / %	± 5	± 10	± 15	–	–

Table 2.11:
Requirements for measurement rooms and weights of a calibration laboratory for mass determination [17, 31].

m mass
R_a, R_z surface roughness (mean roughness value or average roughness depth)
χ magnetic susceptibility
$\mu_0 M_z$ magnetic polarisation
U expanded uncertainty
($U = 2u_c$, see Section 4),
p air pressure
t temperature
h_r relative humidity
dt_{1h}, dt_{12h} maximum temperature change within one or twelve hours
dh_{4h} maximum humidity change within four hours

The temperature dependence of the volume usually cannot be disregarded when the mass standard and the weight to be calibrated have different densities and/or different volume expansion coefficients. For example, this applies to mass comparisons between stainless steel standards and Pt-Ir prototypes. If the volume V of a weight determined using hydrostatic weighing is specified for the reference temperature t_0, the following applies to a temperature t that deviates from t_0:

$$V(t) = V(t_0)\,[1 + \alpha_V(t - t_0)], \qquad (2.25)$$

where:
α_V is the volume expansion coefficient (for the temperature range t_0 to t) and
t_0 is the reference temperature for volume determination (normally, $t_0 = 20$ °C).

The volume expansion coefficients of some materials commonly used in mass determination are listed in Table 2.12.

---

[4] "E_0" is not an official designation for an accuracy class of OIML R 111 [17]. Standards used to calibrate class E_1 weights must have comparable or better metrological characteristics than the weight being calibrated. Thus, the designation "E_0" is a continuation of the system for error limits and uncertainties below the accuracy classes of OIML R 111.

Material	α_V / K^{-1}	Literature
Stainless steel (CrNi 18 8)	4.8×10^{-5}	[32]
Platinum-iridium (90 10)	2.6×10^{-5}	[33]
Brass (CuZn 62 38)	5.4×10^{-5}	[32]
Nickel silver (CuNiZn 62 15 22)	5.4×10^{-5}	[32]
Silicon	7.7×10^{-6}	[34]
Zerodur®	$\leq 1.5 \times 10^{-7}$	[35]

Table 2.12:
Volume expansion coefficients α_V ($t = 20\,°C$) of some materials commonly used in mass determination

Example:

Temperatures: $t_0 = 20\,°C$, $t = 21.5\,°C$

Mass standard (platinum-iridium):

$$m_R = 1\ kg$$
$$V_R(t_0) = 46.4300\ cm^3$$
$$V_R(t) = 46.4318\ cm^3$$
$$\Delta V_R = V_R(t) - V_R(t_0) = 0.0018\ cm^3$$

Specimen (stainless steel):

$$m_T = 1\ kg$$
$$V_T(t_0) = 124.3800\ cm^3$$
$$V_T(t) = 124.3890\ cm^3$$
$$\Delta V_T = V_T(t) - V_T(t_0) = 0.0090\ cm^3$$

Total change of volume difference:

$$\Delta V_T - \Delta V_R = 0.0072\ cm^3$$

Change in air buoyancy: $(\Delta V_T - \Delta V_R) \cdot \rho_a = 8.6\ \mu g$

In this example, the temperature dependence of the volume plays a role that cannot be disregarded since so-called prototype balances, i.e. 1 kg mass comparators with a standard deviation of s ≤ 1 μg, are used for mass comparisons with Pt-Ir prototypes (see Appendix A.3). On the other hand, air buoyancy changes due to volume changes can always be disregarded for mass comparisons between weights made of the same material, e.g. stainless steel or brass, since both the expansion coefficients and the densities are very similar.

2.5.3 Air pressure, relative humidity, adsorption

Air pressure changes during a mass comparison can affect the weighing value, since the latter can only be pressure compensated for a certain density of the weights being calibrated. Many analytical and comparator balances that display the conventional mass are pressure compensated for the reference density $\rho_c = 8000\ kg\ m^{-3}$. If the material density deviates from this value, e.g. in the case of platinum-iridium ($\rho = 21500\ kg\ m^{-3}$), air pressure deviations can cause an increase in the standard deviation for the mass comparison. Therefore, for highest-precision weighing with Pt-Ir prototypes, the prototype balance is frequently installed in a pressure-tight enclosure (see Appendix A.3). High-accuracy mass determinations with analytical and comparator balances are usually carried out in a restricted humidity range between 40 % and 60 %. A relative humidity of less than 40 % may cause electrostatic charges and significant systematic errors. On the other hand, very high relative humidity ($h_r > 60$ %) increases the risk of corrosion. Furthermore, humidity changes always affect the mass standard or the specimen due to sorption effects on the surfaces. While they are exposed to air, the surfaces of solids are always covered with adsorption layers consisting mainly of chemically and physically sorbed water [36, 37]. The chemisorbed layers have considerably larger binding energies than the physisorbed layers, so that the former are almost independent while the latter are dependent on relative humidity. For example, the surface coverage of polished stainless steel standards at a relative humidity of $h_r = 0$ % is between $\mu_{h=0} \leq 0.1$ µg cm^{-2} (cleaned surfaces) and a maximum of approximately $\mu_{h=0} = 0.8$ µg cm^{-2} (uncleaned surfaces or surfaces which have not been cleaned for a long time) [38, 39]. Therefore, the mass of the chemisorbed layer can be estimated at approximately 15 µg to 120 µg for a 1 kg steel standard (surface $A = 150$ cm^2).

The physisorbed adsorption layer on stainless steel standards changes depending on the surface condition, i.e. surface cleanliness and roughness, with the relative humidity according to the Brunauer-Emmett-Teller (BET) equation [38, 40]

$$\mu = \frac{m_A}{A} = \mu_{h=0} + \frac{\mu_m \, c_B \, h}{(1-h)\left[1+(c_B-1)\cdot h\right]}. \qquad (2.26)$$

The parameters of the so-called BET isotherms in equation (2.26) are: m_A mass of the adsorption layer, A surface of the weight, $h = h_r/100$, h_r relative humidity in %, $\mu_{h=0}$ surface coverage for $h = 0$, μ_m change to surface coverage due to a monomolecular layer, c_B BET constant.

For carefully polished stainless steel standards (average roughness depth $R_z \leq 0.1$ µm), the following parameters were determined experimentally [38, 39]:

- Cleaned surfaces ($\mu_h \leq 0.1$ µg cm^{-2}):
  $\mu_m = 0.0084$ µg cm^{-2}
  $c_B = 8.9$
- Uncleaned surfaces ($\mu_{h=0} \geq 0.7$ µg cm^{-2}):
  $\mu_m = 0.018$ µg cm^{-2}
  $c_B = 11.2$

a)

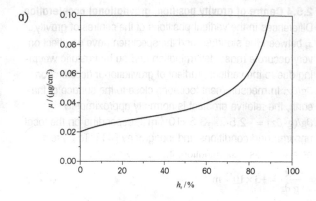

Figure 2.12:
Experimentally determined and inter-
polated BET adsorption isotherms
for polished stainless steel surfaces
(average roughness depth
$R_z \leq 0.1\,\mu m$) [38, 39]:

a) For cleaned surfaces with a sur-
face coverage of $\mu_{h=0} \leq 0.1\,\mu g\,cm^{-2}$.
b) For uncleaned surfaces with a sur-
face coverage of $\mu_{h=0} \geq 0.7\,\mu g\,cm^{-2}$.
μ surface coverage = mass of
      the adsorption layer per unit of
      surface area,
h_r relative humidity in %,
$h = h_r/100$,
$\mu_{h=0}$ surface coverage for $h = 0$.

b)

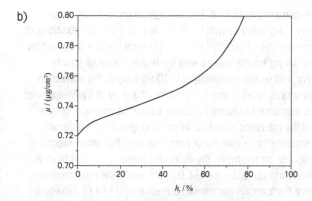

Figure 2.12 shows the curve shapes of the two BET
isotherms. For example, the mass of a cleaned 1 kg stainless steel weight ($A = 150\ cm^2$) increases by $\Delta m_A = 2.9\ \mu g$
and the mass of an uncleaned 1 kg weight increases by
$\Delta m_A = 6.2\ \mu g$ when the relative humidity increases from
20 % to 70 %. This illustrates that such changes in mass
caused by adsorption layers also need to be considered only
in case of the highest-accuracy mass determination (weighing with prototype balances).

In addition to the cleanliness of the surface, roughness can
also influence adsorption behaviour and, therefore, the long-
term stability of weights. International upper limit values
have been established for the average peak-to-valley height
R_z and the average roughness value R_a (arithmetic mean of
the deviations from the centre line of the roughness profile)
for weights [17], which also apply to mass standards used
in accredited calibration laboratories [31] (see Table 2.11).

2.5.4 Centre of gravity position, gravitational acceleration

Differences in the vertical positions of the centres of gravity z_s between the standard and the specimen have an effect on very accurate mass determination and on hydrostatic weighing due to the vertical gradient of gravitational acceleration $\partial g/\partial z$. In measurement locations close to the surface of the earth, the relative gradient is normally approximately $\partial g/(g \cdot \partial z) = -2.5 \dots -3.5 \times 10^{-7}$ m^{-1}, depending on the local underground conditions and topography [41]. To correct precise mass determination, the approximation

$$\frac{\partial g}{g\, \partial z} = -3 \times 10^{-7} \text{ m}^{-1} \tag{2.27}$$

is generally sufficient. For example, a mass comparison of two 1 kg weights with a difference in the vertical positions of the centres of gravity of $\Delta z_s = 20$ mm results in a correction of +6 µg for the weight with the higher centre of gravity. For hydrostatic weighing of a 10 kg weight, the correction amounts to -1.5 mg if the centre of gravity of the weight that is immersed in water is 50 cm below the centre of gravity of the reference standard in air. Changes in gravitational acceleration g that occur over time can also affect weighing results; for example, the daily and monthly relative fluctuations are up to $\Delta g/g = \pm 1.5 \times 10^{-7}$, while the maximum relative fluctuations per minute are $\pm 0.8 \times 10^{-9}$ [41]. However, this only affects proportional weighing with g-dependent (force compensated) analytical balances (Sections 2.3 and 2.4). Since gravitational acceleration also depends on the geographic latitude and altitude, weighing instruments of special accuracy and weighing instruments of high accuracy are adjusted at their place of use (Appendix A.7 and A.8). On the other hand, changes in gravitational acceleration do not play a role in differential weighing.

2.5.5 Electrostatic fields

Electrostatic forces between the weighed object and the weighing instruments and/or the environment can cause significant variations in the instrument indication as well as unknown systematic weighing errors. Whereas the electrical potential of metallic components of an instrument, especially the housing and the suspension, can be brought to the same potential (ground potential), the weighing of non-conductive (dielectric) objects frequently causes problems. The following procedures can be used to discharge or reduce the effect of electric charges:

– Increasing the relative humidity at the measurement location or installing the instruments in a practically closed chamber in which the relative humidity is artificially increased;
– Shielding electrostatic forces by inserting the non-conductive object in a metallic container with a known mass (Faraday cage); since only the conductive surface matters, a thin metallic foil is sufficient;
– Discharging the dielectric body using a suitable radioactive preparation (e.g. α-emitter) or ionised air, generated by a high-voltage discharge.

2.5.6 Magnetic fields

Magnetic fields outside and inside the weighing instruments (e.g. for instruments with electromagnetic force compensation) can also cause systematic weighing errors if the magnetic susceptibility of the weighed object is too high or if it is itself magnetised [17, 42].

Therefore, international limit values have been specified for the permanent magnetic polarisation $\mu_0 M_z$ (magnetisation M_z) and the magnetic susceptibility χ of weights (see Table 2.4 and Table 2.5); exceeding these values is not recommended [17]. If the magnetic polarisation and susceptibility of a weight are smaller than the specified maximum values, it is assumed that the contribution of the measurement uncertainty resulting from the magnetic characteristics of the weight is small enough to be disregarded when calculating the combined uncertainty of the mass determination. The maximum values for polarisation and susceptibility have, in fact, been selected so that during mass determination of a weight, there is no deviation larger than $1/10$ of the maximum permissible error [17, 24].

The magnetic properties of a mass standard should be determined prior to calibration in order to ensure that their influence on the mass determination can be disregarded. A corresponding investigation for weights made of aluminium is not required since they are not magnetic and their susceptibility is clearly below 0.01. Methods to verify the magnetic properties of weights are described in the OIML recommendation R 111 [17].

2.5.7 Mechanical influences

Mechanical influences on weighing instruments can include vibration or tilting (inclination). The influence of vibration depends on the type of instrument and the direction of the vibration vector. For example, a mechanical balance with equal arms is not very sensitive to vertical movements, moderately sensitive to horizontal movements, and very sensitive to tilting.

On the other hand, all instruments that use other forces to compensate for all or part of the gravitational force – that is, all types of load cells as well as electromagnetically compensated instruments – are susceptible to vertical movements. The frequency range that affects weighing is between the reciprocal value of the time interval for one weighing cycle (approx. 10^{-3} Hz) and the reciprocal value of the shortest averaging period for an indicated value (approx. 10 Hz). To reduce weighing errors due to mechanical influences, installing the instruments on a stable, solid weighing table that stands directly on the floor or is attached to a stable wall is recommended.

Systematic errors due to eccentric loads on the weighing pan occur particularly with top-loading instruments; therefore, special centring facilities are offered for comparator balances (e.g. the "LevelMatic®" or a gimbal-mounted load receptor) which largely eliminate errors caused by eccentric loading.

3. Density and volume determination

3.1 Simple density and volume determination

In order to calculate the air buoyancy correction, the volumes or densities of the mass standards and weights being calibrated must be known. For mass determinations with a relative measurement uncertainty of $U(m)/m \geq 1.5 \times 10^{-6}$, the manufacturer's density specifications are normally sufficient if the following relative uncertainties can be assumed [31]:

- Stainless steel ($m \geq 1$ g): $U_\rho /\rho \leq 0.006$
- Stainless steel (1 mg $\leq m \leq$ 500 mg): $U_\rho /\rho \leq 0.02$
- Nickel silver (10 mg $\leq m \leq$ 500 mg): $U_\rho /\rho \leq 0.02$
- Aluminium (1 mg $\leq m \leq$ 10 mg): $U_\rho /\rho \leq 0.05$

If a relative uncertainty of $U(m)/m < 1{,}5 \times 10^{-6}$ is required, the density of the weights with a mass $m \geq 1$ g must be determined (e.g. by hydrostatic weighing). Various methods for density determination are described in the international recommendation OIML R 111 [17].

3.2 Volume determination using hydrostatic weighing

3.2.1 The hydrostatic weighing procedure

Two fundamental principles are used for hydrostatic weighing to determine density [7, 17, 43]:

a) Measuring the reduction in weight of a solid in a liquid;

b) Measuring the increase in weight (of a container with liquid) after submerging a solid.

Since the level of uncertainty that can be achieved with method a) is lower, it is normally used to determine the density of weights; this method is explained in more detail below.

Prior to hydrostatic weighing of a solid T, both its mass m_T and its volume V_T (or its density ρ_T) are unknown. Therefore, two independent weighings in media with different densities (usually air and distilled water) are typically conducted as shown in Figure 3.1.

The first weighing is a substitution weighing of the test piece T compared to the reference standard R_1 (mass m_{R1}, volume V_{R1}), in air with a density of ρ_{a1}; the weighing equation is:

$$m_T = m_{R1} + \rho_{a1} \cdot (V_T - V_{R1}) + \Delta m_{W1}. \tag{3.1}$$

M. Borys et al., *Fundamentals of Mass Determination*,
DOI: 10.1007/978-3-642-11937-8_3, © Springer-Verlag Berlin Heidelberg 2012

Figure 3.1:
Principle of hydrostatic weighing to determine the density of a test piece T with the mass standards R_1 and R_2.

a Substitution weighing in air.
b Substitution weighing in a liquid (e.g. distilled water).

with $\Delta m_{W1} = m_{W1T} - m_{W1R}$ being the difference of the weighing values (balance indications) for T and R_1. Here – and in the following – small weight differences Δm_{Wi} are assumed so that a correction according to equation (2.13) or (2.14) is not required.

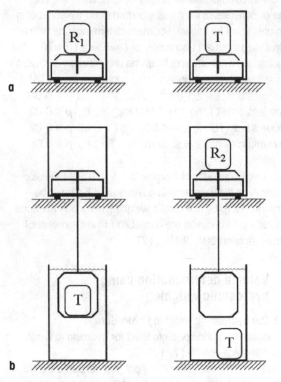

The second weighing is a substitution weighing of the test piece T in a liquid with the density ρ_F compared to the standard R_2 (usually a combination of several standards) in air of the density ρ_{a2}; in this case, the weighing equation is

$$m_T = m_{R2} - \rho_{a2} \cdot V_{R2} + \rho_F \cdot V_T + \Delta m_{W2} \qquad (3.2)$$

with $\Delta m_{W2} = m_{W2T} - m_{W2R}$ as the difference of the weighing values for T and R_2. The volume of the solid is determined from the equations (3.1) and (3.2) to be

$$V_T = \frac{(m_{R1} - \rho_{a1} \cdot V_{R1} + \Delta m_{W1}) - (m_{R2} - \rho_{a2} \cdot V_{R2} + \Delta m_{W2})}{\rho_F - \rho_{a1}} .$$

$$(3.3)$$

3.2.2 Density of water

For hydrostatic weighing, distilled water is normally used, since it has the lowest volume expansion coefficient of all liquids coming into consideration ($\alpha_{v,20\,°C} = 2.1 \times 10^{-4}\ K^{-1}$) [32, 43]. The density ρ_W of distilled water that is free of air can be calculated very precisely as a function of the water temperature t_W ($U(\rho_W)/\rho_W < 10^{-5}$) [44, 47]; a sample table of calculated water densities is found in [43]. For the temperature range from 0 °C to 40 °C and air pressure of 101325 Pa, [44] specifies the function:

$$\rho_W = a_5 \cdot \left[1 - \frac{(t_W + a_1)^2 \cdot (t_W + a_2)}{a_3 \cdot (t_W + a_4)} \right] \tag{3.4}$$

with: $a_1\ /\ °C = -3.983035$
$a_2\ /\ °C = 301.797$
$a_3\ /\ °C^2 = 522528.9$
$a_4\ /\ °C = 69.34881$
$a_5\ /\ (kg\ m^{-3}) = 999.972$

Here the water temperature t_W must be specified in °C in order to obtain the water density ρ_W in kg m^{-3}. The value for a_5 refers to conventional tap water. Due to the influence of isotopic abundance, $a_5 = 999.97495$ (kg m^{-3}) applies to standard mean ocean water (SMOW) [44]. In the temperature range $19.5\,°C \leq t_W \leq 20.5\,°C$, the following simplified approximation[5] applies:

$$\rho_W = 998{,}203 - 0{,}206 \cdot (t_W - 20). \tag{3.5}$$

3.2.3 Influencing quantities in hydrostatic weighing

This section will address the influencing quantities and factors that are essential during hydrostatic weighing:

- Water temperature
- Suitability of the instruments
- Suitability of the wire for weighing with a suspension below the balance
- Wire immersion depth and water surface tension
- Vertical gradient of the gravitational acceleration
- Air bubbles when submerging the specimen in the liquid

---

[5] The relative deviations of this approximation to equation (3.4) are $< 1 \times 10^{-6}$ in the specified temperature range.

Hydrostatic weighing is normally carried out at a water temperature of t_W which is as close as possible to the reference temperature $t_0 = 20\ °C$ in order to avoid uncertainties caused by corrections due to volume expansion.

Weighing instruments that have corresponding openings in the instrument housing and/or the weighing chamber to mount a suspension for weighings below the balance, are especially suitable for hydrostatic weighing. A wire or thread with a bulge at its immersion point – which is included when weighing – connects the weighing pan to the suspension below the balance. The mass of the bulge is proportional to the diameter of the wire. Since the bulge cannot be reproduced easily, the wire diameter and therefore the bulge mass is kept as small as possible. Platinum-iridium wires have proven practical; at diameters of 0.07 / 0.1 / 0.15 / 0.2 / 0.5 mm, they have a load capacity of 100 / 250 / 500 / 800 / 5000 g [41, 43]. Thin tear-proof synthetic threads are also used frequently. A suspension mechanism or a basket to hold the solid inside the liquid can be attached to such a wire or thread.

Varying wire immersion depths can cause errors, especially when older mechanical balances are used. In this case, the sum of the standards R_2 for weighing in water should be selected so that the resulting weighing difference Δm_{W2} is as small as possible. During empty weighing (suspension below the balance with no load, weighing pan of the balance loaded with R_2), the test piece T remains in the liquid, for example, by being placed on a storage shelf in the liquid reservoir.

If there is a large height difference between R_2 and T, the vertical gradient of gravitational acceleration $\partial g/(g \cdot \partial z)$ must be considered in the weighing equation (3.2) (see Section 2.5.4). For instance, the correction for 10 kg weights is 3 mg per metre height difference.

When the specimen is submerged in the liquid, any air bubbles adhering to the surfaces must be carefully removed (e.g. with a plastic-coated wire) in order to avoid systematic errors caused by increased buoyancy.

3.3 Density and volume comparators

For the purpose of volume determination according to equation (3.3), the density of the liquid ρ_F is the reference density. If the hydrostatic weighing is performed in comparison with a density or volume standard that has a known mass and density (or volume), information about the density of the liquid is not required. However, it is necessary to ensure that changes in the density of the liquid during the weighing procedure are small enough so that they can be disregarded. Substitution weighing with a test piece with the mass m_T and the volume V_T in the liquid, and substitution standards with the mass m_{ST} and a volume V_{ST} in air with the density ρ_{aT} (see Figure 3.1b) leads to the weighing equation

$$m_T - \rho_F V_T = m_{ST} - \rho_{aT} V_{ST} + \Delta m'_{WT} . \tag{3.6}$$

For substitution weighing of the volume standard with the volume V_R and the mass m_R in the liquid and substitution standards with the mass m_{SR} and the volume V_{SR} in air with the density ρ_{aR}, the following applies in the same way

$$m_R - \rho_F V_R = m_{SR} - \rho_{aR} V_{SR} + \Delta m'_{WR} . \tag{3.7}$$

From equations (3.6) and (3.7), the volume of the test piece is calculated as

$$V_T = V_R \frac{m_T - m_{ST} + \rho_{aT} V_{ST} - \Delta m'_{WT}}{m_R - m_{SR} + \rho_{aR} V_{SR} - \Delta m'_{WR}} . \tag{3.8}$$

To determine the density of the specimen ρ_T from the reference density of the standard ρ_R, equation (3.8) results in

$$\rho_T = \rho_R \frac{1 - (k_R m_{SR} + \Delta m'_{WR}) / m_R}{1 - (k_T m_{ST} + \Delta m'_{WT}) / m_T} , \tag{3.9}$$

with $k_R = 1 - \rho_{aR} / \rho_{SR}$ and $k_T = 1 - \rho_{aT} / \rho_{ST}$.

Since the density or volume specifications generally refer to the reference temperature $t_0 = 20\,°C$, their changes for a given temperature t must be considered with the aid of the volume expansion coefficient α_V in the form

$$V(t) = V(t_0)[1 + \alpha_V (t - t_0)] \tag{3.10}$$

or

$$\rho(t) = \frac{\rho(t_0)}{1 + \alpha_V (t - t_0)} . \tag{3.11}$$

Suitable density or volume standards are cubes, cylinders or spheres made of stainless steel, Zerodur, quartz or silicon with a sufficiently low density or volume uncertainty.

To meet the highest requirements, national metrology institutes currently use silicon spheres as primary standards. The mass and the volume of these spheres are directly traceable to the SI base units the metre and the kilogram. The density of the silicon spheres used by PTB as the primary standards for the density of solids has been determined with a relative standard uncertainty of approximately 1×10^{-7} [45, 46]. Since the mass of the sphere was determined with a smaller uncertainty than the volume, the uncertainty of volume determination dominates the uncertainty budget.

Figure 3.2:
Example of a volume comparator designed for the density and volume determination of OIML weights and spheres.
(Volume Comparator VC1005X, Mettler-Toledo)

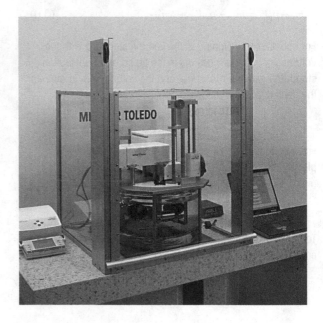

4. Measurement uncertainty

4.1 Terminology

The terminology used here conforms to the "Guide to the Expression of Uncertainty in Measurement" (GUM) [48] that has been published by ISO on behalf of various international organisations (BIPM, IEC, IFCC, ISO, IUPAC, IUPAP, OIML). The same terminology is used in the OIML recommendation R 111 [17], which applies the general rules of the GUM to the calculation of the uncertainties of weights of classes E_1 to M_3. According to the GUM, the following important basic terms must be distinguished:

Standard uncertainty
In many cases the measurand Y is not measured directly, but is determined from N input quantities $X_1, X_2, ..., X_N$ with the aid of a function f:

$$Y = f(X_1, X_2, ..., X_N). \tag{4.1}$$

The input quantities X_i are not exactly known, but are available as estimates x_i (e.g. as mean values from repeated measurements). With the function f, the estimate of the output quantity (measurand) is obtained analogue to equation (4.1):

$$y = f(x_1, x_2, ..., x_N). \tag{4.2}$$

The uncertainty of the measurand y is determined from the standard uncertainties $u(x_i)$ of the input quantities x_i. The standard uncertainties $u(x_i)$ are determined by two different methods known as Type A and Type B.
Type A denotes a method of evaluation by the statistical analysis of series of observations (standard deviation, variance). In the following, a Type A standard uncertainty is referred to as u_A.
Type B denotes a method of evaluation by means other than the statistical analysis of series of observations, e.g. by taking information from a calibration certificate. In the following, a Type B standard uncertainty is referred to as u_B.

Combined standard uncertainty
The standard uncertainty of the result of a measurement, y, is obtained from the standard uncertainties $u(x_1)$, $u(x_2)$,..., $u(x_N)$ of all the input quantities according to the general law of error propagation. If there are correlated input quantities, the respective covariances must be taken into account

M. Borys et al., *Fundamentals of Mass Determination*,
DOI: 10.1007/978-3-642-11937-8_4, © Springer-Verlag Berlin Heidelberg 2012

[48, 49]. The combined standard uncertainty of the output quantity y is denoted by $u_c(y)$.

Expanded uncertainty

The expanded uncertainty U defines an interval about y that may be expected to encompass a large fraction of the distribution of values that could reasonably be attributed to the measurand Y. U is obtained by multiplying the combined standard uncertainty $u_c(y)$ by a coverage factor k:

$$U = k \cdot u_c(y) . \tag{4.3}$$

The result of a measurement is then conveniently expressed as:

$$Y = y \pm U . \tag{4.4}$$

The value of the coverage factor k is chosen on the basis of the level of confidence required. In most cases, it is appropriate to use $k = 2$ which corresponds to a level of confidence of approximately 95% in case of a normal distribution. Whenever an expanded uncertainty U is stated, the selected factor k must also be stated, unless it is known from certain specifications applicable in the respective context. For example, the factor $k = 2$ is generally used by calibration services that are members of the European Cooperation for Accreditation (EA).

4.2 Input quantities and uncertainties in differential weighing

For mass determinations according to the differential weighing method, see equations (2.16) and (2.17), the uncertainty of the mass m_T of the test piece, which is the output quantity y here, is composed of the standard uncertainties of the following input quantities:

– Mass m_R of the reference standard
– Mass m_{Zi} of the auxiliary weights on the reference standard and/or the test piece
– Air density ρ_A
– Volumes V_R and V_T of the reference standard and the test piece
– Adjustment constant c_J (or sensitivity E) of the instrument
– Weighing difference Δm_W (or $\Delta A/E$) (the approximation $\Delta m_W' = \Delta m_W$ is admissible here)

In the following, the calculation of the standard uncertainties $u(x_i)$ of all input quantities and the calculation of the combined standard uncertainty $u_c(m_T)$ and the expanded uncertainty $U(m_T)$ of the mass m_T will be explained.

Standard uncertainty of the weighing difference (Type A)

The weighing difference Δm_W (or $\Delta A/E$, in units of mass) is calculated from n repeated weighing cycles RTTR; the associated standard uncertainty is of Type A

$$u_A^2 = s_n^2 = \frac{s^2}{n} \, ,$$ (4.5)

where s is the empirical standard deviation of a single measurement value Δm_{Wi}

$$s^2 = \frac{1}{n-1} \sum_{i=1}^{n} \left(\Delta m_{Wi} - \overline{\Delta m_W} \right)^2$$ (4.6)

and s_n is the empirical standard deviation of the mean of the n single values Δm_{Wi}

$$\overline{\Delta m_W} = \frac{1}{n} \sum_{i=1}^{n} \Delta m_{Wi} \, .$$ (4.7)

If the number of weighing cycles, n, is less than five, a "pooled standard deviation" s_P (determined in former measurements) should be used for the standard deviation of the balance instead of the empirical standard deviation according to equation (4.6) [17].

Standard uncertainty of the mass standards (Type B)

The standard uncertainty u_R of the reference mass m_R is calculated from the information in the corresponding certificate by dividing the expanded uncertainty U_R by the specified factor k and combining it with the uncertainty u_{inst} due to the instability of the mass of the reference standard:

$$u_R = \sqrt{\left(\frac{U_R}{k} \right)^2 + u_{inst}^2} \, .$$ (4.8)

The uncertainty contribution due to the instability of the reference standard $u_{inst}(m_R)$ can be estimated from observed mass changes after the reference weight has been calibrated several times. If no prior calibration values are available, the uncertainty must be estimated based on empirical values. According to equation (4.8), also the standard uncertainties u_{Zi} of the mass m_{Zi} of possible additional weights on the reference standard or the test piece are calculated. If combinations of several mass standards m_{Ri} or m_{Zi} are used, possible covariances must be taken into account. If these are not specified in a certificate or if no other information is available, it is suitable to calculate the standard uncertainty u_R of the total mass of all standards used as the sum of the standard uncertainties u_{Ri} or u_{Zi}:

$$u_R = \sum u_{Ri} + \sum u_{Zi} \, .$$ (4.9)

Equation (4.9) represents an upper estimate where a correlation coefficient $r = 1$ is assumed.

Standard uncertainty of the air buoyancy correction (Type B)

The standard uncertainty u_b of the air buoyancy correction is calculated from the standard uncertainty of the air density $u_{\rho a}$ (see Appendix A.1) and the standard uncertainties u_{VR} and u_{VT} of the volumes of the standard and the test piece:

$$u_b^2 = (V_T - V_R)^2 \cdot u_{\rho a}^2 + \rho_a^2 \cdot (u_{VT}^2 - u_{VR}^2). \qquad (4.10)$$

The uncertainty of the volume of the auxiliary weights can usually be neglected. The second term in equation (4.10) has the special characteristic that it becomes negative for $u_{VT} < u_{VR}$. This situation can arise, for example, if a mass standard R is used, and its volume or density is only estimated instead of being determined by hydrostatic weighing. The reason for the negative sign then is that the volume of mass standards is usually only determined once, so that the air buoyancy corrections for consecutive mass determinations are correlated. Of course the negative sign only applies if the same (estimated) volume, and hence the same term, was used (with a positive sign) in the preceding calibration of the standard R.

If the densities ρ_R and ρ_T of the standard and the test piece are used instead of the volumes, the following applies instead of equation (4.10):

$$u_b^2 = \left(m_R \cdot \frac{\rho_R - \rho_T}{\rho_R \rho_T} u_{\rho a} \right)^2 + (m_R \rho_a)^2 \cdot \left(\frac{u_{\rho T}^2}{\rho_T^4} - \frac{u_{\rho R}^2}{\rho_R^4} \right). \qquad (4.11)$$

Additional correlations of the air buoyancy corrections must be taken into account if, for example, the reference standard R was calibrated using a primary standard of different density (e.g. a Pt-Ir prototype with volume V_{Pt}) and if, for both mass determinations, the same devices were used to measure the air density. In this case, the following applies instead of equation (4.10) due to the correlation of the air density determinations [28, 49]:

$$u_b^2 = \left[(V_T - V_R)^2 + 2 \cdot (V_T - V_R) \cdot (V_R - V_{Pt}) \right] \cdot u_{\rho a}^2 + \rho_a^2 \cdot (u_{VT}^2 - u_{VR}^2).$$
$$(4.12)$$

Standard uncertainty of the adjustment (or sensitivity) of the balance (Type B)

The standard uncertainty u_s of the adjustment (or sensitivity) of the balance is obtained from the weighing difference Δm_W (or Δ_{AW}/E, in units of mass) and the relative uncertainty u_J/k_J of the adjustment constant k_J or the relative uncertainty u_E/E of the sensitivity E:

$$u_s = \left| \Delta m_W \right| \cdot \frac{u_J}{k_J} \tag{4.13}$$

or

$$u_s = \left| \frac{\Delta A_W}{E} \right| \cdot \frac{u_E}{E} . \tag{4.14}$$

For electromagnetically compensated instruments that are adjusted regularly, $u_J/k_J \leq 5 \times 10^{-4}$ can be assumed. The standard uncertainty u_s can normally be neglected, especially if the weighing difference was reduced by using auxiliary (additional) weights (see Section 2.5.1).

In addition, the Type B uncertainty contributions also include the uncertainties caused by the resolution of the instrument and by eccentric loading [17, 87]. However, this will not be explored in greater detail here since the instrument resolution is usually not the limiting factor and eccentric loading can usually be neglected due to the self-centring effect of modern mass comparators.

Combined standard uncertainty

For mass determination, the Type B standard uncertainty consists of the standard uncertainties u_R, u_b, and u_s:

$$u_B^2 = u_R^2 + u_b^2 + u_s^2 . \tag{4.15}$$

With the Type A standard uncertainties – see equation (4.5) – and the Type B standard uncertainties – see equation (4.15) – the combined standard uncertainty of the specimen is calculated as

$$u_c^2 (m_T) = u_A^2 + u_B^2 . \tag{4.16}$$

Expanded uncertainty

For mass determination, the coverage factor $k = 2$ is normally used to calculate the expanded uncertainty of the test mass m_T [17, 48, 50]:

$$U(m_T) = 2 \cdot u_c (m_T) . \tag{4.17}$$

If an estimate for the standard deviation of the balance is not known, and the number of measurements cannot easily be increased to at least ten, and if the Type A uncertainty u_A is the predominant component of the uncertainty budget – that is, $u_A > u_c(m_T)/2$ – then the factor k should be calculated assuming a confidence level of 95.5 % and the effective degrees of freedom ν_{eff} from the t-distribution [17]. Table 4.1 shows the factor k for different effective degrees of freedom. The effective degrees of freedom associated with the standard measurement uncertainty $u_c(m_T)$ can be estimated with the Welch-Satterthwaite formula [48, 51]:

$$\nu_{eff} = \frac{u_c^4(m_T)}{\sum\limits_{i=1}^{n} \dfrac{u_i^4}{\nu_i}} \ . \tag{4.18}$$

Here u_i are the individual uncertainty contributions to the combined standard uncertainty which result from the measurement uncertainties that are associated with the values x_i of the input quantities that have been assumed to be statistically independent. ν_i are the effective degrees of freedom of the uncertainty contributions u_i [48]. In the case $\nu_i \to \infty$ equation (4.18) is simplified to [17]

$$\nu_{eff} = (n-1)\frac{u_c^4(m_T)}{u_A^4} \ . \tag{4.19}$$

In legal metrology in particular, the conventional mass m_{cT} of weights is often used instead of the mass m_T. The uncertainty of the conventional mass is calculated by analogy with the procedure described above, except for the following differences:
(i) Instead of the standard uncertainties of the masses u_{Ri} and u_{Zi} – see equation (4.9) – the corresponding uncertainties of the conventional masses u_{cRi} or u_{cZi} must be used:

$$u_R' = \sum u_{cRi} + \sum u_{cZi} \tag{4.20}$$

ν_{eff}	1	2	3	4	5	6	7	8	10	20	50	∞
k	13.97	4.53	3.31	2.87	2.65	2.52	2.43	2.37	2.28	2.13	2.05	2.00

Table 4.1:
Coverage factor k for various effective degrees of freedom ν_{eff}

(ii) Instead of the standard uncertainty u_b of the air buoyancy correction – see equation (4.10) – u_b' must be used:

$$u_b'^2 = (V_T - V_R)^2 \cdot u_{\rho_a}^2 + (\rho_a - \rho_0)^2 \cdot (u_{VT}^2 + u_{VR}^2) \qquad (4.21)$$

$\rho_0 = 1.2$ kg m^{-3} is the conventionally defined reference air density. Equation (4.21) is an approximation that covers almost all practical applications; the exact equation can be found in [17].

Because of the second term on the right-hand side of equation (4.21), $(u_R'^2 + u_b'^2) \leq (u_R^2 + u_b^2)$ is always valid. That is the reason why the combined standard uncertainty of the conventional mass $u_c(m_{cT})$ is always smaller than or equal to that of the mass $u_c(m_T)$.

4.3 Example – Calculating mass and mass uncertainty

The mass m_T (and the conventional mass m_{cT}) of a weight of class F$_1$ (nominal value 500 g) is to be determined on an electronic analytical balance ($Max = 1$ kg, $d = 0.1$ mg) using a reference standard of class E$_2$ (mass m_R, volume V_R), so that the expanded uncertainty $U(m_T)$ or $U(m_{cT})$ does not exceed $1/3$ of the maximum permissible error of class F$_1$, i.e. $U(m_{cT}) \leq U(m_T) \leq 0.83$ mg.

(i) Input quantities and input quantity uncertainty

The following input quantities and uncertainties are known:

Test piece (mass to be determined):

Density:	$\rho_T = 8400$ kg m^{-3}	(manufacturer's specification)
Relative uncertainty:	$U(\rho_T)/\rho_T = U(V_T)/V_T = 0.006$	($k = 2$)
Volume:	$V_T = 59.52$ cm^3	(calculated with ρ_T)
Standard uncertainty:	$u_{VT} = 0.18$ cm^3	(calculated)

Standard:

Mass:	$m_R = 500.00045$ g	(calibration certificate)
Expanded uncertainty:	$U(m_R) = 0.75$ mg	($k = 2$)
Standard uncertainty:	$u_{R1} = 0.375$ mg	(calculated[4])
Conventional mass:	$m_{cR} = 499.99950$ g	(calibration certificate)
Expanded uncertainty:	$U(m_{cR}) = 0.25$ mg	($k = 2$)
Standard uncertainty:	$u_{cR1} = 0.125$ mg	(calculated[4])
Density:	$\rho_R = 7900$ kg m^{-3}	(manufacturer's specification)
Relative uncertainty:	$U(\rho_R)/\rho_R = U(V_R)/V_R = 0.006$	($k = 2$)
Volume:	$V_R = 63.29$ cm^3	(calibration certificate)
Standard uncertainty:	$u_{VR} = 0.19$ cm^3	(calculated)

Auxiliary standards:

Mass, conventional mass:	$m_{ZR} = m_{cZR} = 2.003$ mg	(calibration certificate)
Expanded uncertainty:	$U(m_{Z1}) = U(m_{cZ1}) = 0.002$ mg	($k = 2$)
Standard uncertainty:	$u_{Z1} = u_{cZ1} = 0.001$ mg	(calculated[6])
Volume:	$V_{ZR} = 0.0008$ cm^3	(calibration certificate)
Relative uncertainty:	$U(\rho_{ZR})/\rho_{ZR} = U(V_{ZR})/V_{ZR} = 0.05$	($k = 2$)
Standard uncertainty:	$u_{VZR} = 0.00002$ cm^3	(can be neglected)

Air density determination:

Air density:	$\rho_a = 1.2073$ kg m^{-3}	(see Appendix A.1)
Standard uncertainty:	$u_{ra} = 0.0010$ kg m^{-3}	(see Appendix A.1)

Instrument adjustment (or sensitivity):

Relative standard uncertainty: $u_J/k_J \leq 5 \cdot 10^{-4}$ (checked regularly)

Mass determination (with 3 RTTR weighing cycles):

Average weighing difference:	$\Delta m_W = m_{WT} - m_{WR} = +0.09$ mg	($n = 3$)
Standard deviation:	$s = 0.14$ mg	[calculated according to equation (4.6)]

Due to the small number of RTTR cycles ($n < 5$), a "pooled standard deviation" of $s_p = 0.33$ mg is used for the balance that is known from previous measurements.

---

[6] Calculated according to equation (4.8) for a negligibly small contribution of u_{inst}.

(ii) Output quantities and their uncertainties
The following output quantities and uncertainties are calculated using the input quantities:

a) Mass of the test piece:
The weighing equation is as follows:

$$m_T = m_R + m_{ZR} + \rho_a \cdot (V_T - V_R - V_{ZR}) + \Delta m'_W$$

Since $\Delta m_W \ll 3000\ d$, $\Delta m'_W = \Delta m_W$ applies, so that the mass of the test piece is calculated as follows:

$$
\begin{aligned}
m_P &= 500\,000.45\ \text{mg} + 2.003\ \text{mg} + \\
&\quad + 1.2073\ \text{mg cm}^{-3} \\
&\quad (59.52\ \text{cm}^{-3} - 63.29\ \text{cm}^{-3} - 0.0008\ \text{cm}^{-3}) + 0.09\ \text{mg} \\
&= 499\,997.99\ \text{mg} \\
&= 500\ \text{g} - 2.01\ \text{mg}
\end{aligned}
$$

Uncertainty of the mass of the test piece
(i) Type A standard uncertainty:

$$u_A = \frac{s_p}{\sqrt{n}} = \frac{0.33\ \text{mg}}{\sqrt{3}}$$
$$= 0.19\ \text{mg}$$

(ii) Type B standard uncertainty:

$$
\begin{aligned}
u_R &= u_{R1} + u_{Z1} \\
&= 0.375\ \text{mg} + 0.001\ \text{mg} \\
&= 0.376\ \text{mg}
\end{aligned}
$$

$$
\begin{aligned}
u_b^2 &= (V_T - V_R)^2 \cdot u_{\rho a}^2 + \rho_a^2 \cdot (u_{VT}^2 - u_{VR}^2) \\
&= (59.52\ \text{cm}^3 - 63.29\ \text{cm}^3)^2 \cdot (0.0010\ \text{mg cm}^{-3})^2 + \\
&\quad + (1.2073\ \text{mg cm}^{-3})^2 \cdot \left[(0.18\ \text{cm}^3)^2 - (0.19\ \text{cm}^3)^2\right] \\
&= (0.004\ \text{mg})^2 - (0.073\ \text{mg})^2 \\
&= -(0.073\ \text{mg})^2 \qquad \text{(negative sign!)}
\end{aligned}
$$

$$
\begin{aligned}
u_s &= |\Delta m_W| \cdot \frac{u_J}{k_J} \\
&= 0.09\ \text{mg} \cdot 0.0005 \\
&= 0.000045\ \text{mg}
\end{aligned}
$$

$$
\begin{aligned}
u_B &= \sqrt{u_R^2 + u_b^2 + u_s^2} \\
&= \sqrt{(0.376\ \text{mg})^2 - (0.073\ \text{mg})^2 + (0.000045\ \text{mg})^2} \\
&= 0.369\ \text{mg}
\end{aligned}
$$

(iii) Combined standard uncertainty:

$$u_c(m_T) = \sqrt{u_A^2 + u_B^2}$$
$$= \sqrt{(0.19 \text{ mg})^2 + (0.369 \text{ mg})^2}$$
$$= 0.415 \text{ mg}$$

(iv) Expanded uncertainty ($k = 2$):

$$U(m_T) = 2\,u_c(m_T)$$
$$= 0.83 \text{ mg}$$

b) Conventional mass of the test piece

The conventional mass m_{cT} is calculated from the mass m_T according to equation (2.4) as:

$$m_{cT} = 500 \text{ g} + 1.56 \text{ mg}.$$

Uncertainty of the conventional mass of the test piece

(i) Type A standard uncertainty:

$$u_A = \frac{s_p}{\sqrt{n}} = \frac{0.33 \text{ mg}}{\sqrt{3}}$$
$$= 0.19 \text{ mg}$$

(ii) Type B standard uncertainty:

$$u_R' = u_{cR1} + u_{cZ1}$$
$$= 0.125 \text{ mg} + 0.001 \text{ mg}$$
$$= 0.126 \text{ mg}$$

$$u_b'^2 = (V_T - V_R)^2 \cdot u_{pa}^2 + (\rho_a - \rho_0)^2 \cdot (u_{VT}^2 - u_{VR}^2)$$
$$= (59.52 \text{ cm}^3 - 63.29 \text{ cm}^3)^2 \cdot (0.0010 \text{ mg cm}^{-3})^2 +$$
$$+ (1.2073 \text{ mg cm}^{-3} - 1.2 \text{ mg cm}^{-3})^2 \cdot \left[(0.18 \text{ cm}^3)^2 - (0.19 \text{ cm}^3)^2\right]$$
$$= (0.004 \text{ mg})^2 + (0.002 \text{ mg})^2$$
$$= (0.0045 \text{ mg})^2$$

$$u_s = |\Delta m_W| \cdot \frac{u_J}{k_J}$$
$$= 0.09 \text{ mg} \cdot 0.0005$$
$$= 0.000045 \text{ mg}$$

$$u_B' = \sqrt{u_R'^2 + u_b'^2 + u_s^2}$$
$$= \sqrt{(0.126 \text{ mg})^2 + (0.0045 \text{ mg})^2 + (0.000045 \text{ mg})^2}$$
$$= 0.126 \text{ mg}$$

(iii) Combined standard uncertainty:

$$u_c(m_{cT}) = \sqrt{u_A^2 + u_B'^2}$$
$$= \sqrt{(0.19\ \text{mg})^2 + (0.126\ \text{mg})^2}$$
$$= 0.230\ \text{mg}$$

(iv) Expanded uncertainty ($k = 2$):

$$U(m_{cT}) = 2\,u_c(m_{cT})$$
$$= 0.46\ \text{mg}$$

Therefore, the relation $U(m_{cT}) \leq U(m_T) \leq 0.83$ mg applies, i.e. the expanded uncertainties $U(m_T)$ and $U(m_{cT})$ are not greater than $1/3$ of the maximum permissible error of class F_1. As already explained in Section 4.2, the uncertainty of the conventional mass is always less than or equal to the uncertainty of the mass. In this example, $U(m_{cT}) \approx U(m_T)/2$ because the uncertainty contribution u_R' is significantly smaller than u_R.

5. Practical recommendations

Modern analytical balances or verifiable weighing instruments of special and high accuracy have been perfected to such an extent that special training for the operator and special weighing rooms are usually not required. Technological progress has made it possible to make operation much simpler, significantly shortening the weighing time, and making instruments so adaptable that they can now be integrated directly into a production process.

However, these advancements bear the risk that disturbing environmental factors may not receive sufficient attention. These factors normally consist of physical effects that are measurable for balances with a high sensitivity and must not be suppressed by them, since they represent actual changes in mass (e.g. slow evaporation, absorption of moisture), or of forces that act on the weighed object and on the weighing pan (e.g. magnetism, static electricity, air flow) and, thereby, negatively affect the instrument indication.

Place of installation

The accuracy and reliability of the weighing results are closely related to the location of the instruments. Points to be considered are listed below.

Weighing table
- Should transmit vibration as little as possible
- Must not bow or sag while working (e.g. sturdy laboratory table, laboratory bench, stone table)
- Must not be magnetic (no steel tabletop or stone plates with reinforcement bars)
- Must be protected against static electricity (no plastic or glass)
- Must stand on the floor or be attached to the wall, but not both at the same time (prevents simultaneous transmission of vibrations from the wall and the floor)
- Must be reserved for the use of weighing instruments

Working environment
- Must be low-vibration
- Must only have one entrance (air currents)
- Must have as few windows as possible (risk of direct solar radiation)
- Corners of the room should be reserved for weighing tables, since these are the most rigid portions of a building with the lowest level of vibration.

M. Borys et al., *Fundamentals of Mass Determination*,
DOI: 10.1007/978-3-642-11937-8_5, © Springer-Verlag Berlin Heidelberg 2012

Temperature

- The room temperature should be kept as constant as possible in order to avoid a temperature drift of the weighing results (typically 1–2 ppm/°C/h).
- Do not weigh in the vicinity of radiators.

Relative humidity

- The relative humidity should be between 40% and 60%.
- The change rate of humidity over time should be kept as small as possible to prevent transients.
- For microbalances, constant monitoring is recommended (correct for changes, if possible).

Light

- Direct solar radiation must be avoided (e.g. preferably a wall without windows).
- Lighting fixtures should be mounted at an adequate distance from the weighing table in order to avoid thermal radiation (especially in case of incandescent lamps; fluorescent lighting is less critical).

Air

- Do not weigh in the vicinity of air conditioners or other devices with fans (e.g. computers).
- Avoid proximity to radiators (the heat source can cause temperature drift and air currents).
- Do not weigh next to a door.

Operating the instruments

Levelling

Check whether the air bubble on the level is centred and correct the position, if required, by adjusting the feet. Then calibrate the instrument.

Wind shield

- In case of an instrument with an adjustable wind shield, set it so that the wind shield opening is minimised.
- For an instrument with a conventional wind shield, only open it far enough so that the weighed object can be placed on the instrument easily (avoid air turbulence and temperature changes).

Switching on

- Always leave the instrument connected and switched on, so that a thermal equilibrium can be established within the instrument.
- Do not turn off the instrument, instead only put it in standby mode. The instrument will then still be supplied with power (no warm-up time required).

Weighing container
- Use the smallest weighing container possible.
- Avoid plastic weighing containers at all times and glass containers at a relative humidity under 30 % – 40 %, since they can become electrostatically charged.
- The weighing container and the goods it contains should be at ambient temperature. Temperature differences cause air currents and changes to the water film on the weighing container and the weighed object.
- Do not use your hands to place the weighing container into the weighing chamber. You could alter the temperature and relative humidity of the weighing chamber and the weighing container, which would be detrimental to the weighing process.

Weighing pan
- Place the weighed object onto the centre of the weighing pan in order to avoid errors caused by eccentric loading.
- In case of micro balances and semi-micro balances that have been idle for an extended period of time ($\geq$ 30 min), a load should first be briefly placed on the weighing pan (initial weighing effect).
- Once the weighing process is complete, remove the weighed object from the weighing pan. This prevents changes to the temperature and relative humidity in the weighing chamber caused by the weighed object.

Reading
- Check if the instrument is exactly at zero before starting the weighing process; tare if required. This prevents zero errors.
- Set the automatic stability detector according to your requirements. Once the instability symbol turns off (= release of the weighing result), read the result immediately.

Calibration/adjustment
- Calibrate/adjust the sensitivity of the instruments regularly, especially the first time you put the instrument into operation, after changing its location, after levelling it, and after significant temperature, humidity, or air pressure changes.
- With instruments that automatically adjust themselves, calibration using an external weight does not need to be carried out as often.

Care
- Keep the weighing chamber and weighing pan clean.
- Use only clean weighing containers.

Physical influences

If the balance indication is not stable, the result increases or decreases slowly, or the values that are displayed are simply wrong, this is frequently caused by undesirable physical influences. The most frequent causes are:
– Improper treatment of the weighed object
– Improper location of the instrument
– Moisture being adsorbed or desorbed by the weighed object
– Electrostatic charges on containers or weighed object
– Magnetic containers or weighed object

Temperature
Effect: The weight displayed for the weighed object changes continuously in one direction.
Corrective action:
– Do not weigh samples taken directly from the dryer or refrigerator
– Acclimatise the weighed object to the temperature of the laboratory/weighing chamber
– Use tongs to hold samples
– Do not reach into the weighing chamber with your hands in order to avoid temperature changes
– Select sample containers with a small surface area

Moisture absorption/evaporation
Effect: The weight of the weighed object increases or decreases continuously.
Cause: You are measuring the weight loss of volatile substances, e.g. the evaporation of water. An increase in weight is caused by weighing hygroscopic goods (moisture is being absorbed from the air). You can easily reproduce this effect by weighing alcohol or silica gel.
Corrective action:
– Use clean, dry weighing containers and keep the load receptor free of dirt and drops of water
– Use containers with a narrow neck
– Use containers with covers
– Use metallic test tube holders instead of cork or cardboard (the latter can absorb or release significant amounts of moisture)

Electrostatics
Effect: Unstable balance indication
Cause: Your sample or weighing container has accumulated an electrostatic charge. Materials with low electrical

conductivity (most glass or plastic weighing containers) can accumulate an electrostatic charge. This charge is mainly caused by contact or friction while handling or transporting materials (especially powders and granulates). If the air is dry (relative humidity less than 40 % = poor surface conductivity), these electrostatic charges do not discharge, or do so only very slowly. Weighing errors are caused by electrostatic forces acting between the weighed object and the environment. If the weighed object and the environment have equal electric charges (+,+ or -,-), they repel each other; if the charges are opposite (+,- or -,+), they attract each other. These electrostatic forces can be measured by weighing instruments of special accuracy and even by precision balances (weighing instruments of high accuracy, see Section 3.4), and lead to the weighing errors described above. A plastic container that was rubbed with a woollen cloth or paper towel exhibits precisely this effect.

Corrective action:

The electrostatic charge of the weighed object must be discharged or shielded. The following options are available:

– Increase the relative humidity using an evaporator or corresponding air conditioner setting. This is especially important in heated rooms during the winter (recommended relative humidity range: 40 % – 60 %).

– Shield electrostatic forces (place the weighing container inside a metal container).

– Use different containers:

Plastic	Glass	Metal
poor	moderate	very good

– Use ionised air to discharge the weighed object (ioniser).

– Electrically ground the instrument (and therefore the weighing pan; with many instruments, this is done automatically via the grounded plug).

Magnetism

Effect: Poor reproducibility of the weighing values, different weighing results depending on the position of the weighed object.

Cause: You are weighing a magnetic (i.e., magnetically permeable or magnetised) material. Magnetic materials attract or repel each other. These resulting additional forces are incorrectly interpreted as a load.

Corrective action:

- Demagnetise ferromagnetic goods (iron, steel, nickel) as much as possible.
- Since force decreases as distance increases, underlay a non-magnetic spacer to place the weighed object further from the weighing pan (e.g. a beaker or aluminium stand, but do not use absorptive materials such as cork matting); the same effect can be achieved with a feed-through for a suspension.
- Use non-magnetic materials whenever possible.
- Shield magnetic forces (place the weighed object into a magnetically permeable metal container)

Static buoyancy

Effect: Different weighing values when weighing at different air densities or in vacuum.

Cause: The force caused by the buoyancy of a body in a medium corresponds to the weight force of the medium displaced by the body (Archimedes principle).

Corrective action:

Since weighing instruments are usually calibrated or adjusted using weights that have a density of 8.0 g/cm³, weighing objects with other densities means that the effect of changes in air buoyancy must be estimated – taking into account air temperature, relative humidity, and air pressure – and the air buoyancy correction must be considered if necessary (see Section 2.5.1).

Gravitation

Effect: The instrument indication changes when, for example, weighing is performed 10 m higher up (e.g. moving from the 1st floor to the 4th floor within a building).

Cause: To determine the mass or the conventional mass, the weighing instrument measures the force of attraction between the earth and the weighed object (gravitation). This force mainly depends on the geographic latitude of the installation location and on the elevation above sea level (see equation A.7.1).

Corrective action:

- Adjust/calibrate the weighing instrument every time it is relocated.
- Use instruments with fully automatic (motor-driven) self-adjustment.

6. Literature

1. Bureau International des Poids et Mesures (BIPM): International Vocabulary of Metrology – Basic and general concepts and associated terms (VIM), 3rd edition, Paris/Sèvres 2010

2. International Organisation of Legal Metrology (OIML): International vocabulary of terms in legal metrology. Paris 2000

3. Bureau International des Poids et Mesures (BIPM): Le Système international d'unités (SI) – The International System of Units (SI). 8th edition, Paris/Sèvres 2006

4. Gläser, M.: 100 Jahre Kilogrammprototyp. PTB-Bericht MA-15, Physikalisch-Technische Bundesanstalt, Braunschweig 1989

5. L'aventure du mètre. Catalogue of the exhibition "Musèe National des Techniques", CNAM, Paris 1989

6. Moreau, H.: Le système Métrique. Chiron, Paris 1975

7. Kochsiek, M.; Gläser, M. (eds.): Comprehensive mass metrology. WILEY-VCH Verlag, Berlin 2000

8. Davis, R. S.: Possible new definition of the kilogram. Phil. Trans. R. Soc. A 363 (2005), pp. 2249–64

9. BIPM 2010 CCM Report of the 12th meeting (2010) to the CIPM Recommendation G1 (2010), to be published on http://www.bipm.org

10. Davis, R. S.: Measurement of mass and density. In: Gläser, M.; Kochsiek, M. (Ed.): Handbook of metrology. Weinheim: WILEY-VCH Verlag 2010, pp. 137–155

11. Gläser, M.; Borys, M.; Ratschko, D.; Schwartz, R.: Redefinition of the kilogram and the impact on its future dissemination. Metrologia 47 (2010), pp. 419–428

12. Recommendation OIML R 76-1: Nonautomatic weighing instruments, Part 1: Metrological and technical requirements – Tests. OIML, Paris 2006

13. Mohr, P. J.; Taylor, B. N.: CODATA values of the fundamental constants: 2006. J. Phys. Chem. Ref. Data 37 (2008), pp. 1187–1284

14. Newton, I.: Philosophiae naturalis principia mathematica. London 1687, facsimile reprint, London 1955

15. DIN 1305: Masse, Wägewert, Kraft, Gewichtskraft, Gewicht, Last. Beuth Verlag, Berlin 1988

16. Girard, G.: The third periodic verification of national prototypes of the kilogram (1988–1992). Metrologia 31 (1994), pp. 317–336

17. International Recommendation OIML R 111: Weights of classes E_1, E_2, F_1, F_2, M_1, M_{1-2}, M_2, M_{2-3} and M_3, Part 1: Metrological and technical requirements. OIML, Paris 2004

M. Borys et al., *Fundamentals of Mass Determination*,
DOI: 10.1007/978-3-642-11937-8_6, © Springer-Verlag Berlin Heidelberg 2012

18. Borys, M.; Scholz, F.; Firlus, M.: Realization of the mass scale.
    PTB-Mitteilungen 118 (2008) No. 2 and No. 3 (Special Issue), pp. 10–15
    (http://www.ptb.de/en, see "Publications")

19. Bich, W.: From the SI mass unit to multiples and submultiples: an overview.
    Metrologia 40 (2003), pp. 306–311

20. Council Directive 74/148/EEC of 4 March 1974 on the approximation of the laws of the
    Member States relating to weights of from 1 mg to 50 kg of above-medium accuracy.
    (http://eur-lex.europa.eu)

21. Council Directive 71/317/EEC of 26 July 1971 on the approximation of the laws of the
    Member States relating to 5 to 50 kilogramme medium accuracy rectangular bar weights
    and 1 to 10 kilogramme medium accuracy cylindrical weights. (http://eur-lex.europa.eu)

22. German Verification Ordinance (Appendix 8: Weights),of 12th August 1988
    (Federal Law Gazette I, p. 1657), Fourth Ordinance amending the Verification
    Ordinance of 8 February 2007 (Federal Law Gazette I, p. 70)

23. International Document OIML D 28: Conventional value of the result of weighing in air.
    OIML, Paris 2004

24. Gläser, M.: Magnetic interactions between weights and weighing instruments.
    Meas. Sci. Technol. 12 (2001), pp. 709–715

25. European Standard EN 45501 (1992): Metrological aspects of non-automatic weighing
    instruments

26. Nater, R.; Reichmuth, A; Schwartz, R.; Borys, M.; Zervos, P.: Dictionary of Weighing
    Terms, Springer Berlin 2009

27. Schwartz, R.: Guide to mass determination with high accuracy. PTB Report MA-40,
    Physikalisch-Technische Bundesanstalt, Braunschweig 1995

28. Schwartz, R.; Borys, M.; Scholz, F.: Guide to Mass Determination with High Accuracy.
    PTB Report MA-80e, Physikalisch-Technische Bundesanstalt, Braunschweig 2006

29. Key Comparison Data Base of the BIPM (http://kcdb.bipm.org/appendixC/)

30. Gläser, M.: Change of the apparent mass of weights arising from temperature
    differences. Metrologia 36 (1999), pp. 183–197

31. Gläser, M.: Advices for the calibration of mass standards.
    PTB-Bericht MA-52, Braunschweig 1997

32. Kohlrausch, F.: Praktische Physik, Band 3: Tabellen und Diagramme.
    Teubner, Stuttgart 1986, 23. Auflage

33. Bureau International des Poids et Mesures (BIPM): Certificate for the prototype of the
    kilogram No. 52. 15th June 2010

34. Schödel, R.; Bönsch, G.: Precise interferometric measurements at single crystal silicon
    yielding thermal expansion coefficients from 12 °C to 28 °C and compressibility.
    In: Decker, J. E.; Brown, N. (eds.): Recent developments in traceable dimensional
    measurements. Proc. SPIE 4401 (2001), pp. 54–62

35. Schott AG: ZERODUR® – Zero Expansion Glass Ceramic. Data sheet, Mainz 2011

36. Seah, M. P.; Qiu, J. H.; Cumpson, P. J.; Castle, J. E.: Stability of reference masses II: The effect of environment and cleaning methods on the surfaces of stainless steel and allied materials. Metrologia 31 (1994), pp. 93 –108

37. Davidson, S.: A review of surface contamination and the stability of standard masses. Metrologia 40 (2003), pp. 324–338

38. Schwartz, R.: Untersuchung des Sorptionseinflusses bei Massebestimmungen hoher Genauigkeit durch Wägung und durch Ellipsometrie unter kontrollierten Umgebungsbedingungen. PTB-Bericht MA-29, Braunschweig 1993

39. Schwartz, R.: Precision determination of adsorption layers on stainless steel mass standards by mass comparison and ellipsometry. Part I: Adsorption Isotherms in Air. Metrologia 31 (1994), pp. 117–128

40. Brunauer, S.; Emmett, P. H.; Teller, E.: Adsorption of Gases in Multimolecular Layers. J. Am. Chem. Soc. 60 (1938), pp. 309–319

41. Kohlrausch, F.: Praktische Physik. Band 1, Teubner, Stuttgart 1996, 24. Auflage

42. Davis, R. S.; Gläser, M.: Magnetic properties of weights, their measurements and magnetic interactions between weights and balances. Metrologia 40 (2003), pp. 339–355

43. Spieweck, F.; Bettin, H.: Methoden zur Bestimmung der Dichte von Festkörpern und Flüssigkeiten. PTB-Bericht W-46, Braunschweig 1991

44. Tanaka, M.; Girard, G.; Davis, R. S.; Peuto, A.; Bignell, N.: Recommended table for density of water between 0 °C and 40 °C based on recent experiment reports. Metrologia 38 (2001), pp. 301–309

45. Bettin, H.; Toth, H.: Solid density determination by the pressure-of-flotation method. Meas. Sci. Technol. 17 (2006), pp. 2567–2573

46. Bettin, H.; Borys, M.; Nicolaus, A.: Density: From the Measuring of a Silicon Sphere to Archimedes' Principle. PTB-Mitteilungen 118 (2008) No. 2 and No. 3 (Special Issue), pp. 16-22 (http://www.ptb.de/en, see "Publications")

47. Kell, G. S.: Density, thermal expansivity and compressibility of liquid water from 0 °C to 150 °C: Correlations and tables for atmospheric pressure and saturation reviewed and expressed on 1968 temperature scale. J. Chem. Eng. Data 20 (1975), pp. 97–105

48. JCGM 100 (2008): The evaluation of measurement data – Guide to the Expression of Uncertainty in Measurement GUM 1995 with minor corrections. http://www.bipm.org/en/publications/guides/gum.html

49. Schwartz, R.: Realization of the PTB's mass scale from 1 mg to 10 kg. PTB Report MA-21e, Braunschweig 1991

50. EA-4/02: Expression of the uncertainty of measurement in calibration. European Cooperation for Accreditation (EA), 1999

51. DAkkS-DKD-3: Angabe der Messunsicherheit bei Kalibrierungen. Deutsche Akkreditierungsstelle GmbH (DAkkS), Braunschweig 2010

52. Davis, R. S.: Equation for the determination of the density of moist air (1981/91). Metrologia 29 (1992), pp. 67–70

53. Picard, A.; Davis, R. S.; Gläser, M.; Fujii, K.: Revised formula for the density of moist air (CIPM-2007). Metrologia 45 (2008) 149–155

54. Preston-Thomas, H.: The International Temperature Scale of 1990 (ITS-90). Metrologia 27 (1990), pp. 3–10

55. Picard, A.; Fang, H.: Three methods of determining the density of moist air during mass comparisons. Metrologia 39 (2002), pp. 31–40

56. Picard, A.; Fang, H.; Gläser, M.: Discrepancies in air density determination between the thermodynamic formula and a gravimetric method: Evidence for a new value of the mole fraction of argon in air. Metrologia 41 (2004), pp. 396–400

57. Gläser, M.; Schwartz, R.; Mecke, M.: Experimental determination of air density using a 1 kg mass comparator in vacuum. Metrologia 28 (1991), pp. 45–50

58. Picard, A.; Fang, H.: Mass comparisons using air buoyancy artefacts. Metrologia 41 (2004), pp. 330–332

59. Kobayashi, Y.: On a more precise correction for air buoyancy and gas adsorption in mass measurement. In: Taylor, B. N.; Phillips, W. D. (Eds.): Precision measurement and fundamental constants II. Nat. Bur. Stand. (U.S.), Spec. Publ. 617 (1984), pp. 441–443

60. Balhorn, R.: Experimentelle Bestimmung der Luftdichte durch Wägung beim Massevergleich. PTB-Mitt. 93 (1983), pp. 303–308

61. Prowse, D. B.: Measurement of air density for high accuracy mass determination. In: Taylor, B. N., Phillips, W. D. (Eds.): Precision measurement and fundamental constants II. Nat. Bur. Stand. (U.S.), Spec. Publ. 617 (1984), pp. 437–439

62. Reichmuth, A.; Richard, P.: Density determination using the Mettler-Toledo M_one mass comparator. Proc. South Yorkshire Int. Weighing Conf. 2003, pp. 15.1–15.13

63. Mizushima, S.; Ueki, M.; Fujii, K.: Mass measurement of 1 kg silicon spheres to establish a density standard. Metrologia 41 (2004), pp. S68–S74

64. Chung, J. W.; Borys, M.; Firlus, M.; Lee, W. G.; Schwartz, R.: Bilateral comparison of buoyancy artefacts between PTB and KRISS. 19th Int. Conf. on Force, Mass & Torque (IMEKO TC3), Cairo 2005, pp. 26.1-26.6 and Measurement 40 (2007), pp. 761–765

65. Borys, M.; Gläser, M.; Mecke, M.: Mass determination of silicon spheres used for the Avogadro project. Proc. 19th Int. Conf. on Force, Mass & Torque IMEKO TC3, Cairo 2005, pp. 13.1–8 and Measurement 40 (2005), pp. 766–71

66. Picard, A.: Mass determinations of a 1 kg silicon sphere for the Avogadro project. Metrologia 43 (2006), pp. 46–52

67. Madec, T.; Meury, P. A.; Sutour, C.; Rabault, T.; Zerbib, S.; Gosset, A.: Determination of the density of air: a comparison of the CIPM thermodynamic formula and the gravimetric method. Metrologia 44 (2007), pp. 441–447

68. Grabe, M.: Note on the application of the method of least squares.
    Metrologia 14 (1978), pp.143–146

69. Cameron, J. M.; Croarkin, M. C.; Raybold, R. C.: Designs for the calibration of standards of mass. Nat. Bur. Stand. (U.S.), Tech. Note 952, 1977

70. Bich, W.: Variances, covariances and restraints in mass metrology.
    Metrologia 27 (1990), pp. 111–116

71. Bich, W.: Bias and optimal linear estimation in comparison calibrations.
    Metrologia 29 (1992), pp. 15–22

72. Zuker, M.; Mihailov, G.; Romanowski, M.: Systematic search for orthogonal systems in the calibrations of submultiples and multiples of the unit of mass.
    Metrologia 16 (1980), pp. 51–54

73. http://en.wikipedia.org/wiki/Metrology

74. http://www.bipm.org

75. http://www.imeko.org

76. http://www.iso.org

77. http://www.iec.ch

78. http://www.euramet.org

79. Schwartz, R.: Internationale Entwicklungen im gesetzlichen Messwesen.
    In: Odin, A.; Peters, M. (Hrsg.): Vertrauen in die Messtechnik – Neue internationale Grundlagen und Richtlinien. PTB-Bericht PTB-Q-3, Braunschweig 2006

80. Directive 2004/22/EC of the European Parliament and of the Council on Measuring Instruments, 31 March 2004, L 135/1 of 30 April 2004

81. http://www.oiml.org

82. http://www.welmec.org

83. Framework for a Mutual Acceptance Arrangement on OIML Type Evaluations (MAA), OIML Document B10-1, Paris 2004

84. http://www.european-accreditation.org

85. http://www.european-accreditation.org/n1/doc/EA-1-08.pdf

86. http://www.ilac.org

87. Guidelines on the Calibration of Non-Automatic Weighing Instruments
    EURAMET/cg-18/v.02 (2009)

88. Moritz, H.: Geodetic reference system 1980. Bulletin Geodesique 54 (1980), pp. 395–405

89. Torge, W.: Geodäsie. Walter de Gruyter, Berlin 2003, 2nd edition

90. WELMEC Guide 2: Directive 90/384/EEC: Common Application, Non-automatic weighing instruments. WELMEC, Issue 5, May 2009, pp. 35–39

91. Schwartz, R., Lindau, A.: Das europäische Gravitationszonenkonzept für eichpflichtige Waagen. PTB-Mitteilungen 113 (2003), No. 1, pp. 35–42

92. Gravity Information System (GIS), http://www.ptb.de/en/org/1/ (see "1.1 Mass")

Appendix

A.1 Determining air density

If an air buoyancy correction according to equations (2.12), (2.16) or (2.17) is required, the air density is calculated using the following parameters of state: Pressure, temperature, relative humidity, and, if applicable, CO_2 content. The most precise values are determined with the internationally recommended thermodynamic formula, the so-called CIPM equation of 2007 [52, 53]. It is based on the equation of state of a non-ideal gas:

$$\rho_a = \frac{p\,M}{Z\,R\,T} \qquad (A.1.1)$$

with:

p air pressure

M molar mass of moist air [see equation (A.1.2)]

Z coefficient of compressibility [see equation (A.1.9)]

R = 8.314472 J mol^{-1} K^{-1} molar gas constant

T = $(273.15 + t)$ thermodynamic temperature in K
   (according to ITS-90 [54])

t temperature in °C

The molar mass M of moist air is calculated from the molar mass M_a of dry air and the molar fraction x_v as well as the molar mass M_v of water vapour in moist air:

$$M = M_a - x_v\left(M_a - M_v\right) \qquad (A.1.2)$$

with: $M_a = [28.96546 + 12.011 \cdot (x_{CO2} - 0.0004)] \cdot 10^{-3}$ kg mol^{-1}
       $M_v = 18.01528 \cdot 10^{-3}$ kg mol^{-1}

The molecular fraction x_{CO2} of CO_2 only needs to be measured when the air density ($u(\rho_a)/\rho_a < 5 \cdot 10^{-4}$) needs to be determined with extremely high accuracy, e.g. when comparing the mass of a 1 kg steel standard with a Pt-Ir kilogram prototype; in all other cases $x_{CO2} = 0.0004$ can be assumed. The molecular fraction x_v of water vapour is not measured directly but is determined using one of the following three methods:

a) Measuring the relative humidity h_r (in %):

$$x_v = \frac{h \cdot f_w(p,t) \cdot p_{sv}(t)}{p} \qquad (A.1.3)$$

with:

$h = h_r/100$, f_w fugacity coefficient of water vapour in air
   [see equation (A.1.8)]

M. Borys et al., *Fundamentals of Mass Determination*,
DOI: 10.1007/978-3-642-11937-8, © Springer-Verlag Berlin Heidelberg 2012

p_{sv} saturation vapour pressure of water
[see equation (A.1.7)]

b) Measuring the dew point temperature t_d:

$$x_v = \frac{f_w(p,t_d) \cdot p_{sv}(t_d)}{p} \qquad \text{(A.1.4)}$$

c) Measuring the temperature difference $(t_t - t_f)$ with an aspiration psychrometer:

$$x_v = \frac{f_w(p,t_f) \cdot p_v(t_t,t_f)}{p} \qquad \text{(A.1.5)}$$

with:

p_v vapour pressure [see equation (A.1.6)]
t_t temperature of the dry thermometer in °C
t_f temperature of the moistened thermometer in °C
The vapour pressure p_v is derived from the saturation vapour pressure p_{sv} according to Sprung's formula [27]:

$$p_v = p_{sv}(t_f) - 6.62 \cdot 10^{-4} \, \text{K}^{-1} \cdot (t_t - t_f) \cdot p \qquad \text{(A.1.6)}$$

For all three measurement methods, the saturation vapour pressure $p_{sv}(t)$ must be calculated according to [52, 53]:

$$p_{sv} = k_1 \exp\left(k_2 T^2 + k_3 T + k_4 + k_5 T^{-1}\right) \qquad \text{(A.1.7)}$$

Equation (A.1.7) is a numerical equation with:
$k_1 = 1$ Pa
$k_2 = 1.2378847 \cdot 10^{-5}$ K^{-2}
$k_3 = -1.9121316 \cdot 10^{-2}$ K^{-1}
$k_4 = 33.93711047$
$k_5 = -6.3431645 \cdot 10^3$ K
T thermodynamic temperature in K

According to [52, 53], the following applies for the fugacity coefficient $f_w(p, t)$:

$$f_w = \alpha + \beta\, p + \gamma\, t^2 \qquad \text{(A.1.8)}$$

with:
$\alpha = 1.00062$
$\beta = 3.14 \cdot 10^{-8}$ Pa^{-1}
$\gamma = 5.6 \cdot 10^{-7}$ K^{-2}
p air pressure in Pa
t temperature in °C

The coefficient of compressibility Z is calculated as follows:

$$Z = 1 - \frac{p}{T}\left[f_1(t) + f_2(t, x_v)\right] + \frac{p^2}{T^2} f_3(x_v) \qquad \text{(A.1.9)}$$

with:
$$f_1(t) = a_0 + a_1 t + a_2 t^2$$
$$f_2(t, x_v) = (b_0 + b_1 t) \cdot x_v + (c_0 + c_1 t) \cdot x_v^2$$
$$f_3(x_v) = d_0 + d_1 x_v^2$$

a_0	$= 1.58123 \times 10^{-6}$ K Pa^{-1}
a_1	$= -2.9331 \times 10^{-8}$ Pa^{-1}
a_2	$= 1.1043 \times 10^{-10}$ K^{-1} Pa^{-1}
b_0	$= 5.707 \times 10^{-6}$ K Pa^{-1}
b_1	$= -2.051 \times 10^{-8}$ Pa^{-1}
c_0	$= 1.9898 \times 10^{-4}$ K Pa^{-1}
c_1	$= -2.376 \times 10^{-6}$ Pa^{-1}
d_0	$= 1.83 \times 10^{-11}$ K^2 Pa^{-2}
d_1	$= -0.765 \times 10^{-8}$ K^2 Pa^{-2}

When applying the CIPM air density formula, the recommended pressure range 600 hPa $\leq p \leq$ 1100 hPa and the recommended temperature range 15 °C $\leq t \leq$ 27 °C must be observed [53]. The relative uncertainty (1σ) of the air density formula CIPM 2007 is $u_f/\rho_a = 2.2 \times 10^{-5}$ without considering the uncertainty of the quantities to be measured [53]. When the maximum possible effort is expended to measure p, t, h_r, and x_{CO2}, a relative standard uncertainty of the air density $u(\rho_a)/\rho_a$ of about 5×10^{-5} is attainable [66, 67].

An even lower measurement uncertainty can be achieved with the aid of experimental determination of the air density, for example, by weighing two buoyancy artefacts [55–62]. The buoyancy artefacts must have approximately the same mass (m_1, m_2), the same surfaces, but different volumes (V_1, V_2). If the mass difference $m_1 - m_2$ was determined by weighing in vacuum, i.e. without an air buoyancy correction, and the volumes have been determined by hydrostatic weighing, the air density is calculated from the weighing difference of the buoyancy artefacts in air, $m_{W1} - m_{W2}$, according to:

$$\rho_a = \frac{(m_1 - m_2) - (m_{W1} - m_{W2})}{V_1 - V_2} \qquad (A.1.10)$$

With this method, relative standard uncertainties of $u(\rho_a)/\rho_a \leq 3 \times 10^{-5}$ have already been achieved for 1 kg buoyancy artefacts [55, 56, 58, 62–67].

The accuracy of the CIPM air density formula is often not required in practice, so that various approximation formulas have been proposed.

With:

x_{CO2} = 0.0004
Z = 0.99961
f_w = 1.0040
p_{SV} = 670.262 Pa $\times$ exp (0.062 $\cdot$ t)
t temperature in °C

the following simple and precise approximation can be derived directly from the CIPM equation for normal environmental conditions, i.e. for 900 hPa $\leq p \leq$ 1100 hPa, 10 °C $\leq t \leq$ 30 °C, $h_r \leq$ 80 % (cf. the recommended approximation in [27], updated for the CIPM equation 2007 [53]):

$$\rho_a = \frac{0.34851 \cdot p - 0.008863 \cdot h_r \cdot \exp(0.062 \cdot t)}{273.15 + t} \quad (A.1.11)$$

In this numerical equation, the pressure has to be inserted in hPa, the relative humidity in %, and the temperature in °C in order to obtain the air density ρ_a in kg m^{-3}. Under the specified environmental conditions, the relative uncertainty of the approximation given by equation (A.1.11) is less than $u_f/\rho_a = 2 \times 10^{-4}$.

In addition to the uncertainty in the formula, u_f, the uncertainties of the measurements for pressure, temperature, and relative humidity determine the uncertainty of the air density ρ_a. With the standard uncertainties u_p, u_T, and u_{hr} of the corresponding measurements, the combined standard uncertainty of the air density $u_c(\rho_a) = u_{pa}$ is:

$$u_{pa} = \sqrt{u_f^2 + \left(\frac{\partial \rho_a}{\partial p} u_p\right)^2 + \left(\frac{\partial \rho_a}{\partial T} u_T\right)^2 + \left(\frac{\partial \rho_a}{\partial h_r} u_{hr}\right)^2}. \quad (A.1.12)$$

Under normal conditions (T = 293.15 K, p = 101325 Pa, h_r = 50 %), the following values can be substituted for the relative sensitivity coefficients in this equation [53]:

$$\frac{\partial \rho_a}{\partial p} = 10^{-5} \text{ Pa}^{-1} \times \rho_a \quad (A.1.13a)$$

$$\frac{\partial \rho_a}{\partial T} = -4 \times 10^{-3} \text{ K}^{-1} \times \rho_a \quad (A.1.13b)$$

$$\frac{\partial \rho_a}{\partial h_r} = -9 \times 10^{-5} \times \rho_a \quad (A.1.13c)$$

An additional uncertainty contribution $(\partial \rho_a / \partial x_{CO2} \times u_{xCO2})^2$ with:

$$\frac{\partial \rho_a}{\partial x_{CO2}} = 0.4 \times \rho_a \quad (A.1.13d)$$

must be considered in equation (A.1.12) when the molar fraction x_{CO2} is also measured in order to satisfy the highest requirements for the determination of air density $(u(\rho_a)/\rho_a < 5 \times 10^{-4})$.

Example
Calculating air density and air density uncertainty
Assume the following measurements and standard uncertainties:

Pressure: $p = 1019.5$ hPa $u_p = 0.5$ hPa
Temperature: $t = 20.05$ °C ($T = 293.20$ K) $u_T = 0.1$ K
Relative humidity: $h_r = 42\%$ ($h = 0.42$) $u_{hr} = 5\%$
CO_2 content: $x_{CO2} = 0.0004$ (assumed value)

Calculating air density according to the CIPM formula:
Saturation vapour pressure $p_{sv}(t)$ according to equation (A.1.7)

$$p_{sv} = 2346.418 \text{ Pa}$$

Fugacity coefficient $f_w(p,t)$ according to equation (A.1.8):

$$f_w = 1.004046$$

Molar fraction x_v of water according to equation (A.1.3):

$$x_v = 0.009705575$$

Coefficient of compressibility according to equation (A.1.9):

$$Z = 0.9996185$$

Molar mass M of moist air according to equation (A.1.2):

$$M = 28.85918 \times 10^{-3} \text{ kg/mol}$$

Air density ρ_a according to equation (A.1.1):

$$\rho_a = 1.207364 \text{ kg/m}^3$$

Calculating air density according to the approximation formula:
Air density ρ'_a according to equation (A.1.11)

$$\rho'_a = 1.207421 \text{ kg/m}^3$$

Therefore, the relative deviation from the CIPM formula is:

$$\left| \frac{\rho'_a - \rho_a}{\rho_a} \right| = 4.7 \times 10^{-5}$$

Calculating the uncertainties:
Absolute uncertainty according to equation (A.1.12):

$$u_{\rho a} = u'_{\rho a} = 0.0010 \text{ kg m}^{-3}$$

Relative uncertainty:

$$\frac{u_{\rho a}}{\rho_a} = \frac{u'_{\rho a}}{\rho'_a} = 8 \times 10^{-4}$$

The standard uncertainties of the air density calculated with the CIPM formula and the approximation are equal in this case, since the uncertainty components u_f of both formulas are small compared to the uncertainties of the measurements.

A.2 Mathematical description of weighing equations

A.2.1 Representation in matrix form

A system of weighing equations is normally represented in matrix form as follows [7, 27, 28, 49]:

$$X \beta = y - e \qquad (A.2.1)$$

$X = (x_{ij})$, $i = 1 \ldots n, j = 1 \ldots k$ is the matrix of the equation system with the coefficients $x_{ij} = +1, -1$ or 0. $\beta = (\beta_j)$ is the vector of the k masses being determined; $y = (y_i)$ is the vector of the n weighing results, that is, the mass differences $\Delta m_i = (m_R - m_T)_i$ determined using one of the equations (2.16) or (2.17); $e = (e_i)$ is the vector of the unknown, random errors of the mass determinations.
Assuming observations with the same accuracy, that is, assuming that the same weighing instrument is used for all mass comparisons and that the weighing results have random, uncorrelated errors, it is possible to show that the error sum of squares $e^T \cdot e$ (e^T = transposed vector of e) is minimised when the condition of regression is met:

$$X^T e = 0 . \qquad (A.2.2)$$

With equation (A.2.1), this leads to the normal equations:

$$X^T X \beta = X^T y . \qquad (A.2.3)$$

Therefore, the expected values of the unknown masses, considering the constraints [see equation (A.2.2)], are:

$$\langle \beta \rangle = L y \qquad (A.2.4a)$$

with:

$$L = \left(X^T X \right)^{-1} X^T . \qquad (A.2.4b)$$

L is frequently referred to as the solution matrix; $(X^T X)^{-1}$ is the inverse of the symmetrical matrix $(X^T X)$.

A.2.2 Consideration of constraints

A characteristic of mass determination is that the observed measurements y_i always represent differences between two masses. As a consequence, the determinant of the matrix of the normal equations equals zero:

$$\text{Det}\left(X^T X\right) = 0 \tag{A.2.5}$$

and an inversion of $(X^T X)$ is not possible without considering constraints. For example, the mass m_R of a reference standard can be taken as a constraint and be described by the following additional equation:

$$\beta_r = m_R, \; r = 1 \dots k \tag{A.2.6}$$

Frequently, the first weight β_1 is used as the reference standard ($r = 1$). There are various approaches to consider one or more constraints: Equation (A.2.6) can either be considered implicitly in vector y, added to the normal equations [equation (A.2.3)], the so-called Lagrange multiplier method [49, 69], or added to the weighing equation system [equation (A.2.1)], the so-called Gauss-Markov approach [70, 71]). For the same constraint, all of these methods return the same solution $\langle \beta \rangle$, but different variance-covariance matrices. Only the Lagrange multiplier method makes it possible to separately calculate the standard uncertainties of Type A and Type B [28, 49].

The initial result of adding the constraint, [equation (A.2.6)], into the system of the normal equations, [equation (A.2.3)], is that the matrix $(X^T X)$ is no longer square and therefore remains non-invertible. The symmetry of $(X^T X)$ can be re-established by adjusting the vector $\langle \beta \rangle$ with a formal parameter, the Lagrange multiplier λ. Therefore, adjusting the normal equations (for $r = 1$) requires the following steps [69]:

- Inserting a row $k+1$ and a column $k+1$ with the coefficients $1,0,\dots,0$ into the matrix $(X^T X)$.
- Supplementing the vector $\langle \beta \rangle$ with the Lagrange multiplier using $\beta_{k+1} = \lambda$.
- Inserting a row $k+1$ and a column $n+1$ with the coefficients $0,\dots,0,1$ into the matrix X^T.
- Supplementing the vector y with the mass of the reference standard using $y_{n+1} = m_R$.

A.2.3 Regression by weighting the weighing equation

When mass determinations are performed with unequal accuracy, that is, with different weighing instruments, the weighing equations must be weighted[7] prior to regression [49]. The matrix W, simply referred to as the weighting matrix below, is used for this purpose. For uncorrelated observations, $W = (w_{ii})$ is a diagonal matrix with the diagonal elements:

$$w_{ii} = \left(\frac{\sigma_0}{s_i}\right)^2, \quad i = 1 \ldots n \tag{A.2.7}$$

s_i is the experimental standard deviation of the mean value of the mass difference $y_i = \Delta m_i$. σ_0 is the normalisation factor with:

$$\sigma_0^2 = \frac{1}{\sum_{i=1}^{n} \frac{1}{s_i^2}} . \tag{A.2.8}$$

which is derived from the standardisation condition

$$\sum_{i=1}^{n} w_{ii} = 1 \tag{A.2.9}$$

Regression is now conducted with the weighted equation system

$$X' = W^{1/2} X \tag{A.2.10a}$$
$$y' = W^{1/2} y \tag{A.2.10b}$$

X' is the weighted matrix of the equation system, and y' is the weighted vector of the weighing results. The normal equations are derived similar to equation (A.2.3)

$$X'^T X' \beta = X'^T y' . \tag{A.2.11}$$

After extending the normal equations for the constraints, the solutions are derived similar to equation (A.2.4a) and (A.2.4b):

$$L' = \left(X'^T X'\right)^{-1} X'^T . \tag{A.2.12}$$

Since the weighting matrix W is a diagonal matrix, the following also applies using the equations (A.2.10a) and (A.2.10b):

$$L = \left(X^T W X\right)^{-1} X^T W . \tag{A.2.13}$$

---

[7] weighting in the mathematical sense, that is, using weighting factors

A.2.4 Calculating uncertainties
with the variance-covariance matrix

According to the general law of error propagation, the variance-covariance matrix of the solution, V_β, is derived from equation (A.2.12a) as follows [27]:

$$V_\beta = L' V_{y'} L'^T \qquad (A.2.14)$$

$V_{y'}$ is the variance-covariance matrix of the weighted weighing equations

$$V_{y'} = \langle e' e'^T \rangle = \sigma^2 I . \qquad (A.2.15)$$

σ^2 is the unknown variance of the weighted observations and I is the unit matrix. The group variance s^2 serves as an estimate for σ^2:

$$s^2 = \langle e'^T e' \rangle = \frac{\sum_{i=1}^{n} \langle e_i' \rangle^2}{v} . \qquad (A.2.16a)$$

$\langle e' \rangle$ is the vector of the weighted residuals $\langle e_i' \rangle$, that is, the estimated values for the unknown errors e'_i, and v is the number of degrees of freedom

$$v = n - k + 1 \qquad (A.2.16b)$$

with:

n number of weighing equations
k number of masses involved.

If the mass differences y_i are determined from n_i repeated RTTR weighing cycles with the standard deviations s_i (empirical standard deviations of the mean values y_i), a better estimate of the group variance is [27]

$$s^2 = \frac{\sum_{i=1}^{n} s_i'^2 + \langle e_i' \rangle^2}{v} , \qquad (A.2.17a)$$

with

$$s_i'^2 = s_i^2 \cdot (n_i - 1) \cdot w_{ii} = \sigma_0^2 \cdot (n_i - 1) \qquad (A.2.17b)$$

and

$$v = \sum_{i=1}^{n} n_i - k + 1 \qquad (A.2.17c)$$

The vector of the weighted residuals $\langle e' \rangle$ is calculated from

$$\langle e' \rangle = y' - \langle y' \rangle \qquad (A.2.18a)$$

with

$$\langle y' \rangle = X' \langle \beta \rangle. \tag{A.2.18b}$$

The vector $\langle y' \rangle$ represents the estimates of the weighted weighing results. To verify the inner consistency of the weighing equations, the unweighted residuals

$$\langle e \rangle = y - \langle y \rangle \tag{A.2.19a}$$

with

$$\langle y \rangle = X \langle \beta \rangle \tag{A.2.19b}$$

are examined, whereby $\langle y \rangle$ is the vector of the corrected (adjusted) weighing results. If a residual e_i is greater than the corresponding standard deviations s_i, this means there must be weighing or evaluation errors during the determination of the mass difference y_i. The ratio of the group standard deviation s according to equation (A.2.16a) or (A.2.17a) to the value σ_0 according to equation (A.2.8) is a measure for the inner consistency of the weighing equations ideally, $s/\sigma_0 = 1$. Values of $s/\sigma_0 > 1.2$ already indicate an excessively large residual and, therefore, inconsistencies in the weighing equations, whereas values of $s/\sigma_0 > 1.5$, usually only result from significant errors (e.g. algebraic sign errors) when entering the weighing equations.

For the variance-covariance matrix of the solutions, equation (A.2.14) with the equations (A.2.12b), (A.2.15), and (A.2.16a) results in:

$$V_\beta = \left(X'^{\mathrm{T}} X' \right)^{-1} X'^{\mathrm{T}} s^2 I \left[\left(X'^{\mathrm{T}} X' \right)^{-1} X'^{\mathrm{T}} \right]^{\mathrm{T}} \tag{A.2.20a}$$

$$= s^2 \left(X'^{\mathrm{T}} X' \right)^{-1} \left(X'^{\mathrm{T}} X' \right) \left(X'^{\mathrm{T}} X' \right)^{-1} \tag{A.2.20b}$$

$$= s^2 \left(X'^{\mathrm{T}} X' \right)^{-1}. \tag{A.2.20c}$$

The variance-covariance matrix V_β is a square, symmetrical matrix with diagonal elements $v_{jj}, j = 1 \dots k$ that represent the variances, and non-diagonal elements $v_{ij}, i \neq j$ that represent the covariances of the mass measurements standards involved.

When applying the Lagrange multiplier method, the variance-covariance matrix is incomplete; that is, it only contains Type A variances and covariances but none of Type B, which would be the case using the Gauss-Markov approach [27]. However, it is for precisely this reason that with the first method the separate calculation of the Type A and Type B standard uncertainties according to the GUM [48] is possible. The uncertainties are then calculated according to the method described in Section 4.2 as explained below [49].

Standard uncertainty of the weighing process (Type A)

With the Lagrange multiplier method, the diagonal elements v_{jj} of the variance-covariance matrix V_β supply the Type A standard uncertainties of the masses β_j directly:

$$u_A^2(\beta_j) = v_{jj}, \; j = 1 \ldots k \tag{A.2.21}$$

Standard uncertainty of the reference standard (Type B)

The standard uncertainty of the reference standard $u_c(m_R)$ is added to the combined standard uncertainty of the masses β_j in the ratio of the nominal values M_j and M_R:

$$u_N(\beta_j) = h_j \, u_c(m_R), \; j = 1 \ldots k \tag{A.2.22a}$$

with

$$h_j = \frac{M_j}{M_R}. \tag{A.2.22b}$$

Standard uncertainty of the air buoyancy correction (Type B)

With the uncertainty of the air density u_{pa} and the volumes of the masses V_j that need to be determined as well as the mass of the reference standard V_R, the standard uncertainty of the air buoyancy correction is derived as follows:

$$u_b^2(\beta_j) = (V_j - h_j \cdot V_R)^2 \cdot u_{pa}^2, \; j = 1 \ldots k. \tag{A.2.23}$$

Equation (A.2.23) applies under the following conditions:
- The air buoyancy corrections of all mass comparisons are correlated (correlation coefficient 1).
- The air densities are equal for all mass comparisons.
- The uncertainties of the volumes u_{Vj} and u_{VR} are small enough that they can be disregarded.

Normally, the standard uncertainties $u_b(\beta_j)$ are also small enough that they can be disregarded, especially if a set of weights with mass standards of the same density is defined within itself.

Standard uncertainty of the sensitivity of the instrument (Type B)

Realizing a mass scale or determining a set of weights within itself requires regular (if possible daily) adjustments or determinations of the sensitivity of the weighing instruments that are used, so that the corresponding standard uncertainty can be assumed to be zero:

$$u_s(\beta_j) = 0, \; j = 1 \ldots k. \tag{A.2.24}$$

Combined standard uncertainty

Similar to equations (4.15) and (4.16), the Type B standard uncertainty is calculated according to:

$$u_B(\beta_j)^2 = u_N(\beta_j)^2 + u_b(\beta_j)^2, \quad j = 1 \ldots k \qquad (A.2.25)$$

and the combined standard uncertainty according to:

$$u_c(\beta_j)^2 = u_A(\beta_j)^2 + u_B(\beta_j)^2, \quad j = 1 \ldots k. \qquad (A.2.26)$$

Expanded uncertainty

With the factor $k = 2$, the expanded uncertainty is derived as:

$$U(\beta_j) = 2\, u_c(\beta_j), \quad j = 1 \ldots k. \qquad (A.2.27)$$

Covariances have to be taken into account if the uncertainty of combinations of several weights must be determined. For example, the variance of the combination of two standards β_1 and β_2 is calculated as:

$$\mathrm{Var}\,(\beta_1 + \beta_2) = \mathrm{Var}\,(\beta_1) + \mathrm{Var}\,(\beta_2) + 2\,\mathrm{Cov}\,(\beta_1, \beta_2)$$
$$\qquad\qquad (A.2.28a)$$
$$= v_{11} + v_{22} + 2\,v_{12} \qquad\qquad (A.2.28b)$$

Covariances can never be disregarded in mass determinations, since they always contain fractions of the uncertainty of the reference standard used. Although the variance-covariance matrix V_β is incomplete when using the Lagrange multiplier method, the standard uncertainties of combinations of mass standards can be correctly specified if the missing Type B variances and covariances that are not included in $V_\beta = (v_{ij})$ are considered as follows:

$$u_c^2(\beta_1 + \beta_2) = u_A^2(\beta_1) + u_A^2(\beta_2) + 2\,v_{12} + [u_B(\beta_1) + u_B(\beta_2)]^2$$
$$\qquad\qquad (A.2.29a)$$
$$= u_A^2(\beta_1) + u_A^2(\beta_2) + 2\,v_{12} + u_B^2(\beta_1) +$$
$$\quad + u_B^2(\beta_2) + 2\,u_B(\beta_1)u_B(\beta_2) \qquad (A.2.29b)$$
$$= u_c^2(\beta_1) + u_c^2(\beta_2) + 2\,[v_{12} + u_B(\beta_1)u_B(\beta_2)]\,.$$
$$\qquad\qquad (A.2.29c)$$

What has just been said applies also in the case of so-called orthogonal weighing schemes [27, 72].

A.2.5 Example of a regression with weighting

A reference standard β_1 (mass $m_R = 1000.0011$ g, expanded uncertainty $U(m_R) = 0.050$ mg, $k = 2$) is used to calibrate the masses β_2 to β_8 (nominal values: $M_2 = 1000$ g, $M_3 = M_4 = 500$ g, $M_5 = M_6 = 200$ g, and $M_7 = M_8 = 100$ g) on a 1 kg mass comparator ($s_p = 0.010$ mg) and a 200 g mass comparator ($s_p = 0.002$ mg). The following weighing scheme has been chosen:

$$
X\beta =
\begin{pmatrix}
+1 & -1 & 0 & 0 & 0 & 0 & 0 & 0 \\
+1 & 0 & -1 & -1 & 0 & 0 & 0 & 0 \\
0 & +1 & -1 & -1 & 0 & 0 & 0 & 0 \\
0 & 0 & +1 & -1 & 0 & 0 & 0 & 0 \\
0 & 0 & +1 & 0 & -1 & -1 & -1 & 0 \\
0 & 0 & 0 & +1 & -1 & -1 & 0 & -1 \\
0 & 0 & 0 & 0 & +1 & -1 & 0 & 0 \\
0 & 0 & 0 & 0 & +1 & 0 & -1 & -1 \\
0 & 0 & 0 & 0 & 0 & +1 & -1 & -1 \\
0 & 0 & 0 & 0 & 0 & 0 & +1 & -1
\end{pmatrix}
\begin{pmatrix}
\beta_1 \\ \beta_2 \\ \beta_3 \\ \beta_4 \\ \beta_5 \\ \beta_6 \\ \beta_7 \\ \beta_8
\end{pmatrix}
$$

$$(A.2.30)$$

In this case the number of weighing equations is $n = 10$, and the number of masses $k = 8$. Let the first six mass comparisons be performed on a 1 kg mass comparator, the other four on a 200 g mass comparator. Each mass comparison shall be repeated six times, that is, six RTTR weighing cycles ($n_i = 6$, $i = 1 \ldots n$) are completed, respectively. The observed differences in mass (mean values) shall be:

$$
y =
\begin{pmatrix}
+0.2031\,(32)\ \text{mg} \\
+0.1984\,(35)\ \text{mg} \\
+0.0035\,(46)\ \text{mg} \\
-0.0972\,(30)\ \text{mg} \\
+0.0061\,(55)\ \text{mg} \\
+0.0455\,(59)\ \text{mg} \\
-0.0495\,(08)\ \text{mg} \\
-0.0006\,(09)\ \text{mg} \\
+0.0509\,(10)\ \text{mg} \\
-0.0496\,(06)\ \text{mg}
\end{pmatrix}
$$

$$(A.2.31)$$

The numbers in brackets specify the empirical standard deviations s_i of the average values y_i in multiples of 0.0001 mg. Equation (A.2.8) and the standard deviations s_i are used to calculate the normalisation factor as

$$\sigma_0 = 0.000379 \text{ mg} , \qquad\qquad (A.2.32)$$

and the weight matrix $W = (w_{ii})$ according to equation (A.2.7) is:

$$W = \begin{pmatrix}
0.0140 & 0 & 0 & 0 & 0 & 0 & 0 & 0 & 0 & 0 \\
0 & 0.0117 & 0 & 0 & 0 & 0 & 0 & 0 & 0 & 0 \\
0 & 0 & 0.0068 & 0 & 0 & 0 & 0 & 0 & 0 & 0 \\
0 & 0 & 0 & 0.0159 & 0 & 0 & 0 & 0 & 0 & 0 \\
0 & 0 & 0 & 0 & 0.0047 & 0 & 0 & 0 & 0 & 0 \\
0 & 0 & 0 & 0 & 0 & 0.0041 & 0 & 0 & 0 & 0 \\
0 & 0 & 0 & 0 & 0 & 0 & 0.2240 & 0 & 0 & 0 \\
0 & 0 & 0 & 0 & 0 & 0 & 0 & 0.1770 & 0 & 0 \\
0 & 0 & 0 & 0 & 0 & 0 & 0 & 0 & 0.1434 & 0 \\
0 & 0 & 0 & 0 & 0 & 0 & 0 & 0 & 0 & 0.3983
\end{pmatrix}$$

$$\qquad\qquad (A.2.33)$$

According to equations (A.2.10a) and (A.2.10b), and with $W^{1/2} = \left(w_{ii}^{1/2}\right)$, the weighted equation system is:

$$X' \beta = y' - e' \qquad\qquad (A.2.34)$$

with

$$X' = \begin{pmatrix}
+0.1183 & -0.1183 & 0 & 0 & 0 & 0 & 0 & 0 \\
+0.1082 & 0 & -0.1082 & -0.1082 & 0 & 0 & 0 & 0 \\
0 & +0.0823 & -0.0823 & -0.0823 & 0 & 0 & 0 & 0 \\
0 & 0 & +0.1262 & -0.1262 & 0 & 0 & 0 & 0 \\
0 & 0 & +0.0688 & 0 & -0.0688 & -0.0688 & -0.0688 & 0 \\
0 & 0 & 0 & +0.0642 & -0.0642 & -0.0642 & 0 & -0.0642 \\
0 & 0 & 0 & 0 & +0.4733 & -0.4733 & 0 & 0 \\
0 & 0 & 0 & 0 & +0.4207 & 0 & -0.4207 & -0.4207 \\
0 & 0 & 0 & 0 & 0 & +0.3787 & -0.3787 & -0.3787 \\
0 & 0 & 0 & 0 & 0 & 0 & +0.6311 & -0.6311
\end{pmatrix}$$

$$\qquad\qquad (A.2.35)$$

and

$$y' = \begin{pmatrix}
+0.02403 \text{ mg} \\
+0.02146 \text{ mg} \\
+0.00029 \text{ mg} \\
-0.01227 \text{ mg} \\
+0.00042 \text{ mg} \\
+0.00292 \text{ mg} \\
-0.02343 \text{ mg} \\
-0.00025 \text{ mg} \\
+0.01927 \text{ mg} \\
0.03130
\end{pmatrix} \qquad\qquad (A.2.36)$$

As described in Section A.2.2, the normal equations (A.2.11) are extended by the constraint $\beta_1 = m_R$ and the Lagrange multiplier λ as follows:

$$\left(X'^{\mathrm{T}} X'\right)\beta = \begin{pmatrix} +0.0257 & -0.0140 & -0.0117 & -0.0117 & 0 & 0 & 0 & 0 & 1 \\ -0.0140 & +0.0208 & -0.0068 & -0.0068 & 0 & 0 & 0 & 0 & 0 \\ -0.0117 & -0.0068 & +0.0392 & +0.0026 & -0.0047 & -0.0047 & -0.0047 & 0 & 0 \\ -0.0117 & -0.0068 & +0.0026 & +0.0385 & -0.0041 & -0.0041 & 0 & -0.0041 & 0 \\ 0 & 0 & -0.0047 & -0.0041 & +0.4099 & -0.2152 & -0.1723 & -0.1729 & 0 \\ 0 & 0 & -0.0047 & -0.0041 & -0.2152 & +0.3763 & -0.1386 & -0.1393 & 0 \\ 0 & 0 & -0.0047 & 0 & -0.1723 & -0.1386 & +0.7234 & -0.0779 & 0 \\ 0 & 0 & 0 & -0.0041 & -0.1729 & -0.1393 & -0.0779 & +0.7228 & 0 \\ 1 & 0 & 0 & 0 & 0 & 0 & 0 & 0 & 0 \end{pmatrix} \begin{pmatrix} \beta_1 \\ \beta_2 \\ \beta_3 \\ \beta_4 \\ \beta_5 \\ \beta_6 \\ \beta_7 \\ \beta_8 \\ \lambda \end{pmatrix}$$

(A.2.37)

$$\left(X'^{\mathrm{T}} X'\right)\beta = X'^{\mathrm{T}} y' =$$

$$\begin{pmatrix} +0.1183 & +0.1082 & 0 & 0 & 0 & 0 & 0 & 0 & 0 & 0 & 0 \\ -0.1183 & 0 & +0.0823 & 0 & 0 & 0 & 0 & 0 & 0 & 0 & 0 \\ 0 & -0.1082 & -0.0823 & +0.1262 & +0.0689 & 0 & 0 & 0 & 0 & 0 & 0 \\ 0 & -0.1082 & -0.0823 & -0.1262 & 0 & +0.0642 & 0 & 0 & 0 & 0 & 0 \\ 0 & 0 & 0 & 0 & -0.0689 & -0.0642 & +0.4733 & +0.4207 & 0 & 0 & 0 \\ 0 & 0 & 0 & 0 & -0.0689 & -0.0642 & -0.4733 & 0 & +0.3787 & 0 & 0 \\ 0 & 0 & 0 & 0 & -0.0689 & 0 & 0 & -0.4207 & -0.3787 & +0.6311 & 0 \\ 0 & 0 & 0 & 0 & 0 & -0.0642 & 0 & -0.4207 & -0.3787 & -0.6311 & 0 \\ 0 & 0 & 0 & 0 & 0 & 0 & 0 & 0 & 0 & 0 & 1 \end{pmatrix} \begin{pmatrix} +0.02403 \text{ mg} \\ +0.02146 \text{ mg} \\ +0.00029 \text{ mg} \\ -0.01227 \text{ mg} \\ +0.00042 \text{ mg} \\ +0.00292 \text{ mg} \\ -0.02343 \text{ mg} \\ -0.00025 \text{ mg} \\ +0.01927 \text{ mg} \\ -0.03130 \text{ mg} \\ 1000001.1 \text{ mg} \end{pmatrix}$$

(A.2.38)

With the solution matrix L' according to the following equation:

$$L' = \begin{pmatrix} 0 & 0 & 0 & 0 & 0 & 0 & 0 & 0 & 0 & 0 & 1 \\ -6.468 & -2.168 & +2.850 & 0 & 0 & 0 & 0 & 0 & 0 & 0 & 1 \\ -1.186 & -3.325 & -1.705 & +3.482 & +0.878 & -0.942 & 0 & 0 & 0 & +0.096 & 0.5 \\ -1.186 & -3.325 & -1.705 & -3.482 & -0.878 & +0.942 & 0 & 0 & 0 & -0.096 & 0.5 \\ -0.474 & -1.330 & -0.682 & +0.097 & -3.083 & -2.925 & +0.764 & +0.567 & -0.101 & -0.019 & 0.2 \\ -0.474 & -1.330 & -0.682 & +0.097 & -3.083 & -2.925 & -0.797 & -0.054 & +0.589 & -0.019 & 0.2 \\ -0.237 & -0.665 & -0.341 & +0.068 & -1.577 & -1.425 & +0.033 & -0.512 & -0.487 & +0.779 & 0.1 \\ -0.237 & -0.665 & -0.341 & +0.029 & -1.506 & -1.500 & +0.033 & -0.512 & -0.487 & -0.798 & 0.1 \end{pmatrix}$$

(A.2.39)

the solutions $\langle \beta \rangle$ are derived according to equation (A.2.12a):

$$\langle \beta \rangle = L' \, y' = \begin{pmatrix} 1000 \text{ g} + 1.100 \text{ mg} \\ 1000 \text{ g} + 0.899 \text{ mg} \\ 500 \text{ g} + 0.402 \text{ mg} \\ 500 \text{ g} + 0.498 \text{ mg} \\ 200 \text{ g} + 0.149 \text{ mg} \\ 200 \text{ g} + 0.199 \text{ mg} \\ 100 \text{ g} + 0.050 \text{ mg} \\ 100 \text{ g} + 0.100 \text{ mg} \end{pmatrix}$$

$$(A.2.40)$$

With the solutions $\langle \beta \rangle$, the adjusted, weighted weighing results $\langle y' \rangle$ and the weighted residuals $\langle e' \rangle$ are calculated according to equations (A.2.18a) and (A.2.18b) as follows:

$$\langle y' \rangle = X' \langle \beta \rangle = \begin{pmatrix} +0.02381 \text{ mg} \\ +0.02171 \text{ mg} \\ -0.00004 \text{ mg} \\ -0.01214 \text{ mg} \\ +0.00019 \text{ mg} \\ +0.00317 \text{ mg} \\ -0.02368 \text{ mg} \\ +0.00003 \text{ mg} \\ +0.01896 \text{ mg} \\ -0.03133 \text{ mg} \end{pmatrix}$$

$$(A.2.41)$$

$$\langle e' \rangle = y' - \langle y' \rangle = \begin{pmatrix} +0.00023 \text{ mg} \\ -0.00025 \text{ mg} \\ +0.00033 \text{ mg} \\ -0.00013 \text{ mg} \\ +0.00023 \text{ mg} \\ -0.00025 \text{ mg} \\ +0.00025 \text{ mg} \\ -0.00028 \text{ mg} \\ +0.00031 \text{ mg} \\ +0.00003 \text{ mg} \end{pmatrix}$$

$$(A.2.42)$$

Accordingly, the unweighted estimates $\langle y \rangle$ and $\langle g \rangle$ are calculated according to equations (A.2.19a) and (A.2.19b):

$$\langle y \rangle = X \langle \beta \rangle = \begin{pmatrix} +0.2012 \text{ mg} \\ +0.2007 \text{ mg} \\ -0.0005 \text{ mg} \\ -0.0962 \text{ mg} \\ +0.0028 \text{ mg} \\ +0.0493 \text{ mg} \\ -0.0500 \text{ mg} \\ +0.0001 \text{ mg} \\ +0.0501 \text{ mg} \\ -0.0496 \text{ mg} \end{pmatrix} \tag{A.2.43}$$

$$\langle e \rangle = y - \langle y \rangle = \begin{pmatrix} +0.0019 \text{ mg} \\ -0.0023 \text{ mg} \\ +0.0040 \text{ mg} \\ -0.0010 \text{ mg} \\ +0.0033 \text{ mg} \\ -0.0038 \text{ mg} \\ +0.0005 \text{ mg} \\ -0.0007 \text{ mg} \\ +0.0008 \text{ mg} \\ +0.0000 \text{ mg} \end{pmatrix} \tag{A.2.44}$$

This shows that all residuals e_i are smaller than the standard deviations s_i [see equation (A.2.31)]; therefore, the weighing equations apparently contain no systematic errors.
The group standard deviation s is calculated according to equation (A.2.17) with the degree of freedom $v = 53$ [see equation (A.2.16b)], as

$$s = 0.000382 \text{ mg} \tag{A.2.45}$$

The ratio of s to the normalisation factor σ_0 [see equation (A.2.32)],

$$\frac{s}{\sigma_0} = 1.01 \tag{A.2.46}$$

is close to the ideal value of 1, thereby confirming the inner consistency of the weighing equations.

With the group variance $s^2 = 1.46 \cdot 10^{-7} \, \text{mg}^2$ and the inverse matrix $(X'^{\text{T}} X')^{-1}$, the variance-covariance matrix according to equation (A.2.20c) is calculated as follows:

$$V_\beta = 1.46 \cdot 10^{-7} \, \text{mg}^2 \cdot \begin{pmatrix} 0 & 0 & 0 & 0 & 0 & 0 & 0 & 0 \\ 0 & 54.66 & 10.02 & 10.02 & 4.01 & 4.01 & 2.00 & 2.00 \\ 0 & 10.02 & 29.16 & 1.57 & 6.53 & 6.53 & 3.34 & 3.19 \\ 0 & 10.02 & 1.57 & 29.16 & 5.76 & 5.76 & 2.80 & 2.96 \\ 0 & 4.01 & 6.53 & 5.76 & 21.45 & 19.83 & 10.03 & 10.06 \\ 0 & 4.01 & 6.53 & 5.76 & 19.83 & 21.51 & 9.96 & 10.00 \\ 0 & 2.00 & 3.34 & 2.80 & 10.03 & 9.96 & 6.24 & 5.01 \\ 0 & 2.00 & 3.19 & 2.96 & 10.06 & 10.00 & 5.01 & 6.27 \end{pmatrix}$$

$$(A.2.47)$$

The uncertainties of the masses β_j are calculated from the diagonal elements v_{jj} of the matrix V_β as well as the combined standard uncertainty of the reference standard, $u_c(m_R) = U(m_R)/2$, according to equation (A.2.4), as follows:

Type A standard uncertainties:

$$u_A(\beta_j) = (v_{jj}^{1/2}) = \begin{pmatrix} 0.0000 \, \text{mg} \\ 0.0028 \, \text{mg} \\ 0.0021 \, \text{mg} \\ 0.0021 \, \text{mg} \\ 0.0018 \, \text{mg} \\ 0.0018 \, \text{mg} \\ 0.0010 \, \text{mg} \\ 0.0010 \, \text{mg} \end{pmatrix}$$

$$(A.2.48)$$

Type B standard uncertainties:
The standard uncertainties $u_b(\beta_j)$ and $u_s(\beta_j)$ [see equations (A.2.23) and (A.2.24)] are small enough so they can be disregarded; therefore, according to equations (A.2.22a), (A.2.22b), and (A.2.25), the following applies:

$$u_B(\beta_j) = u_N(\beta_j) = (h_j \, u_c(m_R)) = \begin{pmatrix} 0.0250 \, \text{mg} \\ 0.0250 \, \text{mg} \\ 0.0125 \, \text{mg} \\ 0.0125 \, \text{mg} \\ 0.0050 \, \text{mg} \\ 0.0050 \, \text{mg} \\ 0.0025 \, \text{mg} \\ 0.0025 \, \text{mg} \end{pmatrix}$$

$$(A.2.49)$$

Combined standard uncertainties:

$$u_c\left(\beta_j\right)=\left(u_A^2\left(\beta_j\right)+u_B^2\left(\beta_j\right)\right)^{1/2}=\begin{pmatrix}0.0250\ \text{mg}\\0.0252\ \text{mg}\\0.0127\ \text{mg}\\0.0127\ \text{mg}\\0.0053\ \text{mg}\\0.0053\ \text{mg}\\0.0027\ \text{mg}\\0.0027\ \text{mg}\end{pmatrix} \qquad \text{(A.2.50)}$$

Expanded uncertainties:

$$U\left(\beta_j\right)=2\,u_c\left(\beta_j\right)=\begin{pmatrix}0.0500\ \text{mg}\\0.0504\ \text{mg}\\0.0254\ \text{mg}\\0.0254\ \text{mg}\\0.0106\ \text{mg}\\0.0106\ \text{mg}\\0.0054\ \text{mg}\\0.0054\ \text{mg}\end{pmatrix} \qquad \text{(A.2.51)}$$

Thus, the overall result of the calibration for the set of weights is:

$$\langle\beta\rangle\pm U\left(\beta_j\right)=\begin{pmatrix}1000.001100\ \text{g}\pm0.050\ \text{mg}\\1000.000899\ \text{g}\pm0.051\ \text{mg}\\500.000402\ \text{g}\pm0.026\ \text{mg}\\500.000498\ \text{g}\pm0.026\ \text{mg}\\200.000149\ \text{g}\pm0.011\ \text{mg}\\200.000199\ \text{g}\pm0.011\ \text{mg}\\100.000050\ \text{g}\pm0.006\ \text{mg}\\100.000100\ \text{g}\pm0.006\ \text{mg}\end{pmatrix} \qquad \text{(A.2.52)}$$

The 500 g weights β_3 and β_4 will be used to illustrate the significance of the covariances. According to equation (A.2.29c), the expanded uncertainty of the sum of both 500 g weights is:

$$\begin{aligned}U\left(\beta_3+\beta_4\right)&=2\,u_c\left(\beta_3+\beta_4\right)\\&=2\sqrt{u_c^2\left(\beta_3\right)+u_c^2\left(\beta_4\right)+2\left[v_{34}+u_B\left(\beta_3\right)u_B\left(\beta_4\right)\right]}\\&=2\sqrt{2\left(0.0127^2\ \text{mg}^2\right)+2\left(1.57\cdot1.46\cdot10^{-7}\ \text{mg}^2+0.0125^2\ \text{mg}^2\right)}\\&=0.0504\ \text{mg}\end{aligned}$$

$$\text{(A.2.53)}$$

Thus, it is identical to the expanded uncertainty $U(\beta_2)$ of the 1 kg mass standard β_2 which makes sense. If the covariances $(v_{34} + u_B(\beta_3)\, u_B(\beta_4))$ were neglected, the expanded uncertainty would be $U(\beta_3 + \beta_4) = 0.0359$ mg. This would be a considerable underestimation.

A.3 Examples of high-resolution mass comparators

The term mass comparator, or comparator balance, has become commonly accepted for high-accuracy weighing instruments used to compare a calibration standard (or test weight) with a reference standard, using the substitution weighing method. The greatest advancements and the smallest (relative) standard deviations in weighing technology have been achieved with 1 kg comparator balances. The highest-accuracy mass determination uses mass comparators almost exclusively (number of scale intervals $n = Max/d > 5 \times 10^7$) [7]. They are mainly used in laboratories that are high up in the hierarchy for the dissemination of the unit of mass (see Section 2.1) such as the International Bureau of Weights and Measures (Bureau International des Poids et Mesures, BIPM), national metrology institutes such as the Physikalisch-Technische Bundesanstalt (PTB), and certain calibration laboratories such as laboratories that are accredited by the Deutscher Kalibrierdienst/Deutsche Akkreditierungsstelle (DKD/DAkkS) (German Calibration Service).

Almost all modern mass comparators are electronic comparator balances with fixed counterweights (see Section 2.4). Most of the weight force is compensated by a lever so that the remaining difference of the weight forces can be measured very accurately (up to $n = 10^{10}$) with electromagnetic (electrodynamic) force compensation.

1 kg mass comparators, which are used for the special purpose of linking the mass of secondary or reference standards to a national prototype of the kilogram, are also called prototype balances. Prototype balances are almost exclusively installed in pressure-tight enclosures, which can usually also be evacuated to carry out precise air density measurements with special buoyancy artefacts or direct mass comparisons under vacuum conditions [55–67]. Figure A.3.1 shows the interior of a 1 kg vacuum mass comparator (prototype balance Sartorius CCL1007); Figure A.3.2 shows another vacuum mass comparator also used by PTB as a prototype balance (Mettler-Toledo M_one).

Table A.3.1 provides an overview of the mass comparators and comparator balances used at PTB for the realisation of the mass scale in the range from 1 mg to 5000 kg and for the calibration of mass standards and weights of the highest quality (accuracy classes "E_0", E_1, and E_2) [18, 27, 49].

Figure A.3.1:
Interior view of a 1 kg vacuum mass comparator with weighing unit and automatic weight-exchange mechanism for eight positions (prototype balance Sartorius CCL1007)

Figure A.3.2:
Prototype balance (Mettler-Toledo M_one) installed in a vacuum chamber (shown open). The mass comparator has an automatic weight-exchange mechanism with six positions.

Range of nominal values	Max / d	Weighing principle	s	s_{rel}
1 mg ... 5 g	6 g / 0.1 µg	Mass comparator with full electromagnetic force compensation	0.3 µg	5×10^{-8}
10 g ... 100 g	111 g / 0.1 µg	Mass comparator with automatic weight-exchange mechanism, 4 positions	0.8 µg	7.2×10^{-9}
100 g ... 1 kg	1 kg / 0.1 µg	Vacuum mass comparator with automatic weight-exchange mechanism, 6 or 8 positions (prototype balance)	0.3 µg	3×10^{-10}
2 kg ... 10 kg	10 kg / 10 µg	Mass comparator with automatic weight-exchange mechanism, 4 positions	20 µg	2×10^{-9}
20 kg ... 50 kg	64 kg / 0.1 mg	Mass comparator with automatic weight-exchange mechanism, 4 positions	0.4 mg	6.3×10^{-9}
100 kg ... 500 kg	500 kg / 0.1 g	Electromechanical balance with automatic weight-exchange mechanism, 2 positions	0.2 g	4×10^{-7}
500 kg ... 5000 kg	5000 kg / 20 mg	Mechanical balance with equal arms and automatic acquisition of measured data	0.6 g	1.2×10^{-7}

Table A.3.1:
Data for the balances and mass comparators (selection) used at PTB to realise the mass scale and for highest-accuracy mass determination.

Max maximum capacity
d scale interval
s standard deviation
s_{rel} relative standard deviation in relation to the usable maximum capacity

A.4 Metrology structures and organisation

Metrology, the science of measurement, can be divided into three areas; all areas include specific tasks and have their own organisational structure:

- scientific metrology, also called fundamental metrology
- legal metrology
- industrial or applied metrology

A.4.1 Scientific metrology

Scientific metrology, also called fundamental metrology, concerns the establishment of quantity systems, unit systems, units of measurement, the development of new measurement methods, the realisation of measurement standards and the transfer of traceability from these standards to users in society [73]. Scientific metrology includes all research activities related to quantities and units of measurement:

- Establishment and further development of the International System of Units (SI)
- Realisation of SI units by appropriate measurement standards and devices
- Dissemination of SI units, i.e. transfer of traceability, from these standards to users in industry and society
- Development of new measurement methods and procedures of measurement, including uncertainty calculation

At an international level, the main player in that field of metrology is the International Bureau of Weights and Measures (BIPM) [74]. It performs its tasks with the authority of the Convention of the Metre, a diplomatic treaty between 54 Member States. Besides the BIPM there are, in an expanded sense, other players in that field such as the International Measurement Confederation (IMEKO) [75], the International Organization for Standardization (ISO) [76] and the International Electrotechnical Commission (IEC) [77].

The institutions of the BIPM (Convention of the Metre) are shown in Figure A.4.1. Its executive institution is the General Conference on Weights and Measures (CGPM) which normally meets every 4 years. It first convened in 1889 and, among other things, approved the international kilogram prototype as the definition of the unit of mass which is still in effect today. CGPM decisions are prepared by the International Committee for Weights and Measures (CIPM). It consists of 18 experts from various member states of the Convention of the Metre, and is held annually at the BIPM. The BIPM operates through a series of Consultative Committees

(CCs), whose members are representatives of the national metrology institutes (NMIs) of the Member States, such as the NIST (USA), AIST/NMIJ (Japan), PTB (Germany), NIM (China), NPL (UK), LNE (France), and METAS (Switzerland). The CCM (Comité consultatif pour la masse et les grandeurs apparentées), for example, is responsible for the physical quantities mass, force, density, pressure, fluid flow, gravimetry, hardness and viscosity. The BIPM itself carries out laboratory work and measurement-related research. It takes part in and organises international comparisons of national measurement standards, and it carries out calibrations for Member States [74]. In 1999, the directors of the NMIs of 38 member states of the BIPM signed the CIPM MRA [74], an agreement for the mutual recognition of the national standards and the calibration and test certificates issued by the NMIs. In the meantime, the CIPM MRA has been signed by the representatives of 82 institutes – from 48 Member States, 31 Associates of the CGPM, and 3 international organisations – and covers a further 135 institutes designated by the signatory bodies [74].

The essential responsibilities of the NMIs are:
- Basic research and development for the representation, maintenance, and dissemination of the units in the International System of Units (SI)
- Research and development in the field of precision measurement technology, especially to determine fundamental and natural constants
- Highest-accuracy calibration and testing, both in the area of industrial metrology and in legal metrology
- National and international committee work, participation in the development of technical regulations and standards
- Advising science, industry, officials, and associations

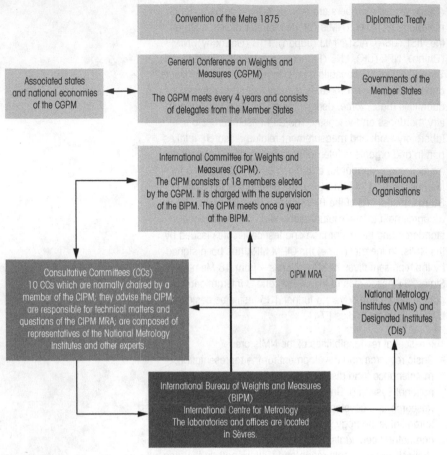

| Convention of the Metre 1875 | ⟷ | Diplomatic Treaty |

| General Conference on Weights and Measures (CGPM)

The CGPM meets every 4 years and consists of delegates from the Member States | ⟷ | Governments of the Member States |

| Associated states and national economies of the CGPM | ⟷ | |

| International Committee for Weights and Measures (CIPM).
The CIPM consists of 18 members elected by the CGPM. It is charged with the supervision of the BIPM. The CIPM meets once a year at the BIPM. | ⟷ | International Organisations |

Consultative Committees (CCs)
10 CCs which are normally chaired by a member of the CIPM; they advise the CIPM; are responsible for technical matters and questions of the CIPM MRA; are composed of representatives of the National Metrology Institutes and other experts.

CIPM MRA

National Metrology Institutes (NMIs) and Designated Institutes (DIs)

International Bureau of Weights and Measures (BIPM)
International Centre for Metrology
The laboratories and offices are located in Sèvres.

Figure A.4.1:
International coordination of metrology: Institutions of the Metre Convention (source: [74])

In the course of increased international cooperation, several regional metrology organisations (RMOs) have formed over the last few decades in order to coordinate the cooperation of NMIs in that region in such fields as research in metrology, traceability of measurements to the SI units, international recognition of national measurement standards and related Calibration and Measurement Capabilities (CMC) of its members. Through knowledge transfer and cooperation among their members, RMOs also facilitate the development of the national metrology infrastructures. In Europe, for example, the regional metrology organisation is EURAMET (European Association of National Metrology Institutes), a legal entity and voluntary association of currently 36 NMIs from the EU and EFTA [78]. EURAMET was inaugurated in 2007 as the successor of EUROMET (European Collaboration in Measurement Standards) that coordinated European metrology successfully over almost 20 years, based on a Memorandum of Understanding (MoU).

A.4.2 Legal metrology

Legal metrology concerns regulatory requirements of measurements and measuring instruments for the protection of health, public safety, the environment, enabling taxation, the protection of consumers and fair trade [73]. Legal metrology covers all technical and administrative processes that are bindingly established in the legal regulations of a country in order to guarantee the quality of measurements in certain areas. Legal metrology protects the buyer when purchasing measurable products and services, guarantees sufficiently accurate measurement results in commerce, and serves to maintain confidence in official measurements, e.g. in the fields of safety and health in the workplace, environmental protection, and radiation protection [79].

As an example, the fundamental goals and responsibilities of legal metrology in Germany are anchored in the Weights and Measures Act, and the technical and metrological requirements as well as verification processes for certain types of measuring devices are anchored in the Verification Ordinance. These legal regulations do not just contain national regulations, but are increasingly implementing European and international requirements and directives such as the European Measuring Instruments Directive (MID) [80] and OIML recommendations [81] in national law.

The International Organisation of Legal Metrology (OIML) was founded in 1955 with the goal of harmonising legal metrology all over the world by [81]:

– Harmonising regulations and test methods in the member states.
– Developing and publishing documents and recommendations regarding the approval and verification of measuring devices.

Meanwhile, the OIML has 113 member states; 57 full and 56 corresponding members, see Figure A.4.2.

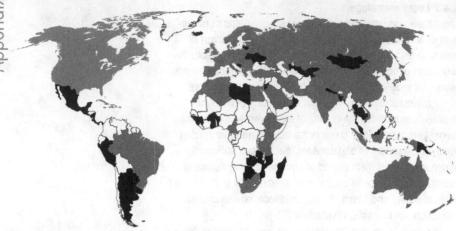

Figure A.4.2:
Member states (blue) and corresponding members (green) of the International Organization of Legal Metrology (OIML)

The structure and institutions of the OIML are illustrated in Figure A.4.3.

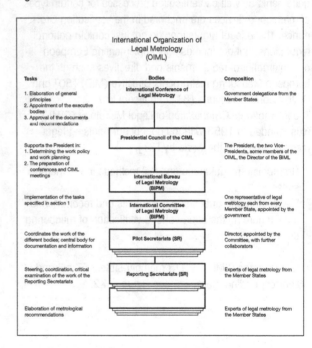

Figure A.4.3:
Structure and institutions of the OIML

In addition to the OIML, there are numerous regional legal metrology organisations (RLMOs) that are responsible for cooperation and harmonisation in the field of legal metrology in their respective region. In Europe, for example, this is WELMEC – the European Cooperation in Legal Metrology [82]. WELMEC was founded in 1990 when a MoU was signed by 18 members consisting of representatives from member states of the European Union (EU) and EFTA.

Upon founding, the acronym WELMEC was formed out of "Western European Legal Metrology Cooperation". However, WELMEC now extends beyond western Europe and also includes member states from central and eastern Europe. Nevertheless, the familiar acronym WELMEC has remained. In May 2004, the 10 new EU member states of Cyprus, the Czech Republic, Estonia, Hungary, Lithuania, Latvia, Malta, Poland, Slovakia, and Slovenia signed the MoU, and Turkey was accepted as an associated member in 2005.

By signing the MoU, the WELMEC member states commit themselves to voluntarily solving noted problems with European legal metrology in a harmonised, consensual manner. In particular, this includes the involvement in WELMEC work groups and the application of published WELMEC guidelines [82].

All industrialised nations have NMIs. In Germany, the Physikalisch-Technische Bundesanstalt (PTB) is notified to test and certify the types of new measuring instruments and devices. But the conformity to type of the serial production of instruments and devices is evaluated and guaranteed by the federal weights and measures authorities or by the manufacturer's quality management system under the supervision of a weights and measures authority.

Just as in scientific metrology, the mutual recognition of test results and certificates is also a central topic in the field of legal metrology. The introduction of the OIML certification system in 1991 represented an important step [81]; it allows the member states to issue certificates confirming that the type (design) of a certain measuring instrument or device conforms with the OIML recommendations. Harmonising requirements, standardising test methods, and recording test results in a specified "Test Report Format" promotes and greatly simplifies the international recognition of test results in the course of national approval processes. This means that a certain type of measuring instrument that has been tested and certified in one country (according to the OIML recommendations) can be certified (approved) in another country without repeating the technical and metrological tests. Since 1991, more than 1500 OIML Certificates of Conformity have been issued for over 40 categories of measuring instruments by about 30 OIML member states for over 300 different manufacturers [81]. So far, the focus has been on certificates for non-automatic weighing instruments, automatic weighing instruments, and load cells. In order to further boost the commitment to mutual recognition, the OIML passed the "Mutual Acceptance Agreement" (MAA) in 2004 [81, 83]. It specifies a framework and precisely defines conditions under which certain member states commit themselves to the binding recognition of test results based on OIML test and evaluation reports. In practice, the MAA is implemented by signing so-called "Declarations of Mutual Confidence" (DoMCs) for each category of measuring instruments (e.g. non-automatic weighing instruments according to OIML R 76).

A.4.3 Industrial metrology

Industrial or applied metrology concerns the application of measurement science to manufacturing and other processes and their use in industry and society, ensuring the suitability of measurement instruments, their calibration and quality control of measurements [73].

Unlike legal metrology, industrial metrology is not controlled by the government but by the needs and requirements of industry and the users of measuring devices. Metrology supplies the information required to analyse, evaluate, and control continuous and discontinuous technical processes in research and production for quality assurance purposes. In order to ensure that the measurements are comparable, reliable, and suitable for verification and approval, the measuring devices and test equipment must be traceable to national and international standards and be calibrated. This is normally done using a hierarchical metrology infrastructure; see Figure A.4.4.

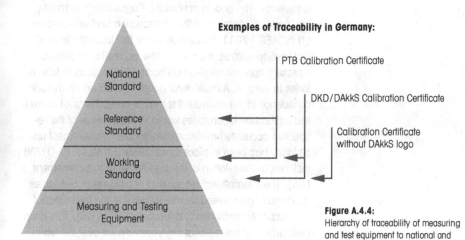

Examples of Traceability in Germany:

PTB Calibration Certificate

DKD/DAkkS Calibration Certificate

Calibration Certificate without DAkkS logo

National Standard

Reference Standard

Working Standard

Measuring and Testing Equipment

Figure A.4.4:
Hierarchy of traceability of measuring and test equipment to national and international standards

The top of this hierarchy is usually formed by an NMI which disseminates the units to commerce, industry, and science with the highest possible level of accuracy by calibrating reference standards. These are followed by subordinate working and production standards and test equipment which are related to the calibrated reference standards; the measurement uncertainty increases with each step. A continuous measuring chain is essential to trace measuring and test equipment to the national standards, which are in turn traced to the International System of Units (SI) using international intercomparison measurements (so-called key comparisons). In this process, the technical and metrological requirements are supplemented with formal requirements, especially the EN ISO 9000 series of standards for QM systems and the EN ISO IEC 17025 standards for test and calibration laboratories. Due to the pronounced increase in the demand for calibration with acknowledged certificates, calibration services were introduced in almost all industrial states. An example is the "Deutscher Kalibrierdienst" (DKD) (German Calibration Service) founded in 1977 that has been under the umbrella of the German Accreditation Service (DAkkS) since 2010. Calibration services provide a market-oriented, metrological infrastructure and are an important link between the NMIs, the manufacturers and users of measuring instruments in commerce and industry. Mean-while, the calibration services are also cooperating closely in regional organisations such as the European Cooperation for Accreditation (EA) [84]. The goal is to achieve equivalency in the way the calibration services work, and the mutual recognition of their certificates. This is done based on mutual confidence and the ongoing exchange of knowledge and experience. The goal is achieved by agreements regarding the mode of operation of the accreditation bodies based on EN ISO/IEC 17011. In addition to the test laboratories and certification centres, the EA member accreditation centres represent approximately 2500 accredited calibration labora-tories in Europe. A multilateral agreement based on mutual confidence in the equivalent technical competence of accred-ited calibration laboratories within the framework of the re-spective accredited minimum assignable measurement un-certainty has been in place since 1989 (EA MLA). EA-01/08 includes a compilation of the signatories of this agreement [85]. The international recognition of calibration certificates is normally guaranteed by membership in the International Laboratory Accreditation Cooperation (ILAC) [86]. This is an association of the regional organisations all over the world. Mutual recognition takes place in a manner similar to the EA, with mutual inspection of individual accreditation centres and comparative measurements.

A.5 Dissemination chain and uncertainties

By definition, the international kilogram prototype has the uncertainty of zero although it, like all solid bodies under atmospheric conditions, is subject to changes in mass due to diffusion, sorption, and similar effects. Currently, the BIPM specifies a standard measurement uncertainty ($k = 1$) of 6 µg for the mass determination of a national prototype of the kilogram. If the national prototype is later used, for example, for the calibration of a 1 kg reference standard with a density of 8000 kg m^{-3}, it cannot be assumed that it still has the mass and uncertainty specified on the BIPM certificate. The mass drift of the prototype and its uncertainty can be estimated from observed mass changes after several recalibrations. If historical data are not available, the estimation has to be based on experience. The current (drift corrected) mass of the prototype is used for the calibration of a mass standard. If, for example, the standard uncertainty of the drift correction is determined to be 3 µg, the standard uncertainty of the mass of the national kilogram prototype at the time of calibration of the mass standard is approximately 6.7 µg. The volume difference between a Pt-Ir prototype and a 1 kg stainless steel standard of approximately 80 cm^3 leads to a buoyancy difference of 96 mg at an air density of 1.2 mg/cm^3. If the air density is determined with, for example, a relative standard uncertainty of 1.2×10^{-4} according to the CIPM air density formula (see Section A.1), this results in the largest contribution to the uncertainty budget, with 11.5 µg. Considering all relevant influences, the mass of the 1 kg stainless steel standard can be determined with a standard measurement uncertainty of approximately 14 µg under these conditions. Numerous national metrology institutes specify minimum standard measurement uncertainties in the range from 15 µg to 25 µg for routine calibrations of 1 kg stainless steel standards (see [29]). Figure A.5.1 provides an overview of the dissemination steps described above, along with typical values for the respective uncertainties. Since they are normally used in certificates, the expanded measurement uncertainties U ($k = 2$) have been specified.

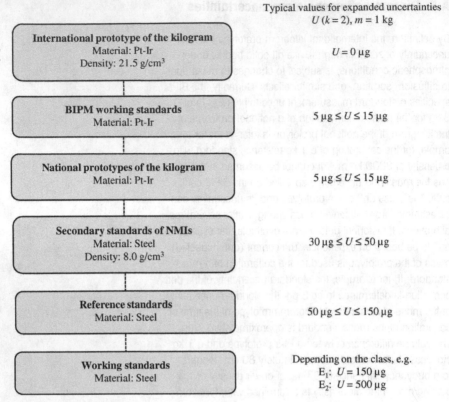

Typical values for expanded uncertainties
$U\ (k = 2),\ m = 1\ \text{kg}$

International prototype of the kilogram
Material: Pt-Ir
Density: 21.5 g/cm³

$U = 0\ \mu g$

BIPM working standards
Material: Pt-Ir

$5\ \mu g \leq U \leq 15\ \mu g$

National prototypes of the kilogram
Material: Pt-Ir

$5\ \mu g \leq U \leq 15\ \mu g$

Secondary standards of NMIs
Material: Steel
Density: 8.0 g/cm³

$30\ \mu g \leq U \leq 50\ \mu g$

Reference standards
Material: Steel

$50\ \mu g \leq U \leq 150\ \mu g$

Working standards
Material: Steel

Depending on the class, e.g.
E_1: $U = 150\ \mu g$
E_2: $U = 500\ \mu g$

Figure A.5.1:
Dissemination chain of the kilogram
with typical values for the expanded
uncertainties $U\ (k = 2)$

A.6 Sample calibration certificate

Physikalisch-Technische Bundesanstalt
Braunschweig und Berlin

Kalibrierschein
Calibration Certificate

Gegenstand:
Object:
1 Massenormal zu 1 kg
1 mass standard of 1 kg

Hersteller:
Manufacturer:
Mettler-Toledo GmbH
Im Langacher
8606 Greifensee

Typ:
Type:
—

Kennnummer:
Serial number:

Auftraggeber:
Applicant:
ROFA Laboratory & Process Analyzers
Hauptstrasse 145
3420 Kritzendorf

Anzahl der Seiten:
Number of pages:
4

Geschäftszeichen:
Reference No.:
1.11-09.061

Kalibrierzeichen:
Calibration mark:
PTB - 06109

Datum der Kalibrierung:
Date of calibration:
16.10.2009

Im Auftrag:
By order:
Braunschweig, 21.10.2009

Siegel
Seal

M. Firlus

Bearbeiter:
Examiner:

M. Hämpke

391 008 j

Kalibrierscheine ohne Unterschrift und Siegel haben keine Gültigkeit. Dieser Kalibrierschein darf nur unverändert weiterverbreitet werden. Auszüge bedürfen der Genehmigung der Physikalisch-Technischen Bundesanstalt.
Calibration certificates without signature and seal are not valid. This calibration certificate may not be reproduced other than in full. Extracts may be taken only with permission of the Physikalisch-Technische Bundesanstalt.

Page 1

Physikalisch-Technische Bundesanstalt

Aufbewahrung
Case

Das Massenormal befindet sich in einem Spezialbehälter aus Kunststoff, der in einem Transport-kasten aus lackiertem Holz untergebracht ist; das Kalibrierzeichen ist auf dem Transportkasten aufge-bracht.
The mass standard is accommodated in a special plastic box, which is inserted in a carriage box of varnished wood; the calibration mark is applied to the carriage box.

Verfahren
Procedure

Die Kalibrierung erfolgte durch Vergleich mit den Hauptnormalen der PTB nach der Substitutions-methode mit Auftriebskorrektur.
The calibration ensued through comparison with the reference standards of PTB using the substitution method with air buoyancy correction.

Die Volumenbestimmung für $m \geq 1$ g erfolgte nach der hydrostatischen Wägemethode.
The volume determination for $m \geq 1$ g was carried out with the hydrostatic weighing method.

Unsicherheit
Uncertainty

Angegeben ist die erweiterte Messunsicherheit, die sich aus der Standardmessunsicherheit durch Multiplikation mit dem Erweiterungsfaktor $k = 2$ ergibt. Sie wurde gemäß dem „Guide to the Expres-sion of Uncertainty in Measurement" (ISO, 1995) ermittelt. Der Wert der Messgröße liegt im Regelfall mit einer Wahrscheinlichkeit von annähernd 95% im zugeordneten Werteintervall.

Die erweiterte Messunsicherheit wurde aus Unsicherheitsanteilen der verwendeten Normale, der Wä-gungen und der Luftauftriebskorrektur berechnet. Eine Abschätzung über Langzeitveränderungen ist in der Unsicherheitsangabe nicht enthalten.

Reported is the expanded uncertainty which results from the standard uncertainty by multiplication with the coverage factor $k = 2$. It has been evaluated according to the "Guide to the Expression of Uncertainty in Measurement" (ISO 1995). Generally, the value of the measuring quantity is found within the attributed interval with a probability of approximately 95%.

The expanded uncertainty was calculated from the contributions of uncertainty originating from the standards used, from the weighings and the air buoyancy corrections. The reported uncertainty does not include an estimate of long-term variations.

Kovarianzen werden im allgemeinen nicht angegeben, daher sind für Kombinationen von Gewichtstü-cken die Unsicherheiten nach der Formel:

$$u_g = \Sigma \; u_i$$

zu addieren. Hierbei sind u_g die Gesamtunsicherheit und u_i die Unsicherheiten der verwendeten Ge-wichtstücke.

Covariances are not generally reported, therefore the uncertainties for combinations of weights must be added according to the above for-mula, with u_g the total uncertainty, u_i the uncertainties of the weights used.

Rekalibrierung
Recalibration

Die Ergebnisse gelten zum Zeitpunkt der Kalibrierung. Es obliegt dem Antragsteller, zu gegebener Zeit eine Rekalibrierung zu veranlassen.
The measurement results are valid at the time of calibration. The applicant is responsible for arranging a recalibration in due course of time.

Page 2

Physikalisch-Technische Bundesanstalt

Umgebungsbedingungen
Ambient conditions

Die Kalibrierung wurde bei folgenden Umgebungsbedingungen ausgeführt:
The calibration was carried out under the following ambient conditions:

	von from	bis to	Unsicherheit uncertainty
Temperatur / °C temperature	21,2	21,3	0,1
rel. Luftfeuchte / % relative humidity	40	41	2
Luftdruck / mbar air pressure	1004,5	1004,6	0,2

Meßergebnisse
Measurement results

Konventioneller Wägewert und Fehlergrenze entsprechend OIML R 111
Conventional mass and maximum permissible error corresponding to OIML R 111

Nennwert nominal value	Kennzeichnung marking	konventioneller Wägewert conventional mass	Unsicherheit uncertainty $k = 2$
1 kg	keine/none	1 kg + 0,13 mg	0,10 mg

Masse
Mass

Nennwert nominal value	Kennzeichnung marking	Masse mass	Unsicherheit uncertainty $k = 2$
1 kg	keine/none	1 kg - 0,11 mg	0,10 mg

Dichte und Volumen
Density and volume

Nennwert nominal value	Kennzeichnung marking	Dichte bei t = 20 °C density at t = 20 °C	Unsicherheit uncertainty $k = 2$	Volumen bei t = 20 °C volume at t = 20 °C	Unsicherheit uncertainty $k = 2$
1 kg	keine/none	8012,44 kgm^{-3}	0,13 kgm^{-3}	124,806 cm^3	0,0020 cm^3

Page 3

105

Physikalisch-Technische Bundesanstalt

Die Physikalisch-Technische Bundesanstalt (PTB) in Braunschweig und Berlin ist das nationale Metrologieinstitut und die technische Oberbehörde der Bundesrepublik Deutschland für das Messwesen und Teile der Sicherheitstechnik. Die PTB gehört zum Dienstbereich des Bundesministeriums für Wirtschaft und Technologie. Sie erfüllt die Anforderungen an Kalibrier- und Prüflaboratorien auf der Grundlage der DIN EN ISO/IEC 17025.

Zentrale Aufgabe der PTB ist es, die gesetzlichen Einheiten in Übereinstimmung mit dem Internationalen Einheitensystem (SI) darzustellen, zu bewahren und – insbesondere im Rahmen des gesetzlichen und industriellen Messwesens – weiterzugeben. Die PTB steht damit an oberster Stelle der metrologischen Hierarchie in Deutschland. Kalibrierscheine der PTB dokumentieren die Rückführung des Kalibriergegenstandes auf nationale Normale.

Dieser Ergebnisbericht ist in Übereinstimmung mit den Kalibrier- und Messmöglichkeiten (CMCs), wie sie im Anhang C des gegenseitigen Abkommens (MRA) des Internationalen Komitees für Maße und Gewichte enthalten sind. Im Rahmen des MRA wird die Gültigkeit der Ergebnisberichte von allen teilnehmenden Instituten für die im Anhang C spezifizierten Messgrößen, Messbereiche und Messunsicherheiten gegenseitig anerkannt (nähere Informationen unter http://www.bipm.org).

The Physikalisch-Technische Bundesanstalt (PTB) in Braunschweig and Berlin is Germany's National Metrology Institute and the supreme technical authority in the Federal Republic of Germany for metrology and certain sectors of safety engineering. The PTB comes under the auspices of the Federal Ministry of Economics and Technology. It meets the requirements for calibration and testing laboratories as defined in EN ISO/IEC 17025.

The central task of the PTB is to realize and maintain the legal units in compliance with the International System of Units (SI) and to disseminate them - in particular within the framework of legal and industrial metrology. The PTB thus is on top of the metrological hierarchy in Germany. The calibration certificates issued by the PTB document that the calibrated object is traceable to national standards.

This certificate is consistent with the Calibration and Measurement Capabilities (CMCs) included in Appendix C of the Mutual Recognition Arrangement (MRA) drawn up by the International Committee for Weights and Measures (CIPM). Under the MRA, all participating institutes recognize the validity of each other's calibration and measurement certificates for the quantities, ranges and measurement uncertainties specified in Appendix C (for details, see http://www.bipm.org).

Physikalisch-Technische Bundesanstalt

Bundesallee 100	Abbestraße 2-12
38116 Braunschweig	10587 Berlin
DEUTSCHLAND	DEUTSCHLAND

Page 4

A.7 Gravity zones in legal metroloy

A close examination of weighing instruments shows that of the four existing accuracy classes, only weighing instruments of class (II) (high accuracy) and class (III) (medium accuracy) need to be addressed as regards gravity zones. The number of scale intervals for instruments of accuracy class (IIII) (ordinary accuracy) is $n \leq 1000$, which means they can be considered insensitive to gravitational acceleration, since the relative changes of indication due to changes of g do not usually exceed 10^{-3} in one country.

On the other hand, instruments of accuracy class (I) (special accuracy) with $n \geq 50000$ always require adjustment with an internal calibration weight or a special external standard weight at the place of use, so that the local g value does not have to be known in this case.

Thus, only weighing instruments of high and medium accuracy need to be addressed as regards gravity zones, because they are sensitive enough to react to gravity changes between the place of adjustment (often the place of manufacturing) and the place of use. In principal, the necessary correction Δm for a mass m can be easily calculated according to the equation

$$\Delta m = m\,(g_2 - g_1)/g_1 \qquad\qquad (A.7.1)$$

if the local g values for the place of adjustment (g_1) and the place of use (g_2) are known. However in practice, it is much easier for a country to establish gravity zones where each zone is assigned a certain g value. The zones are defined so that, depending on the number of scale intervals n of the instrument and the maximum possible relative change in gravitational acceleration $\Delta g/g$, the instrument indication cannot change by more than a specified fraction of the maximum permissible error (mpe). For example, Germany uses the fraction 1, i.e. ± mpe is the criterion. Furthermore, the gravity zones are usually defined in accordance with the administrative boundaries of federal states, counties, or districts of a country. With these two conditions, four gravity zones have been defined for Germany, as shown in Figure A.7.1. A comprehensive survey of gravity information for all European countries is available on the respective website of WELMEC [82].

Figure A.7.1:
Gravity zones and corresponding *g* values defined in Germany

Zone 4
g = 9.8130 m • S⁻²

Zone 3
g = 9.8107 m • S⁻²

Zone 2
g = 9.8081 m • S⁻²

Zone 1
g = 9.8070 m • S⁻²

Depending on the number of scale intervals n, instruments of class (II) and (III) can be adjusted not only for one single zone, but also for two neighbouring zones, or the entire country as shown in Table A.7.1.

Table A.7.1:
Gravity zones in Germany (Z) depending on the number of scale intervals, n, and the accuracy class of a scale

Germany	Zone(s)	g_z in ms^{-2}	Number of scale intervals n	
			Class (II)	Class (III)
Total	Z1-Z4	9.8100	≤ 1000	≤ 3000
Neighbouring zones	Z1-Z2	9.8077		
	Z2-Z3	9.8094	≤ 2000	≤ 5000
	Z3-Z4	9.8118		
Individual zone	Z1	9.8070		
	Z2	9.8081	≤ 3300	≤ 10000
	Z3	9.8107		
	Z4	9.8130		

For example, weighing instruments of medium accuracy (class III) with $n \leq 3000$ and adjusted for $g_z = 9.8100$ m s⁻² can be used all over Germany. Weighing instruments of high accuracy with $n > 1000$ and weighing instruments of medium accuracy with $n > 3000$ must be labelled to indicate the gravitation zone(s) they were adjusted for. Weighing instruments of high accuracy with $n > 3300$ must be adjusted at or for the place of use. The g value for the place of use can be calculated from the geographic latitude φ and the height above sea level h using the equation

$$g = a_1 \left(1 + a_2 \sin^2(\varphi) - a_3 \sin^2(2\varphi)\right) - a_4 h \,. \qquad (A.7.2)$$

with φ geographic latitude

h height above sea level (in m)

$a_1 = 9.780327$ m s^{-2}

$a_2 = 5.3024 \cdot 10^{-3}$

$a_3 = -5.8 \cdot 10^{-6}$

$a_4 = -3.085 \cdot 10^{-6}$ s^{-2}

This formula is derived from the geodetic reference system 1980 (GRS80) [88, 89].

In 1998, WELMEC, the European Cooperation in the field of legal metrology, passed a uniform, optional European gravity zone concept for weighing instruments under legal control [90, 91] which is recognised by nearly all WELMEC member states as an alternative to existing, or a substitute for non-existent, national regulations; see Figure A.7.2.

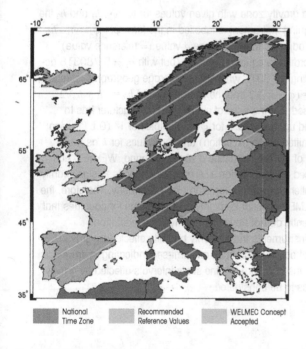

Figure A.7.2:
European countries with national gravity zone regulations and acceptance of the WELMEC gravity zone concept

| National Time Zone | Recommended Reference Values | WELMEC Concept Accepted |

The WELMEC gravity zone concept is independent of political and/or administrative boundaries. A gravity zone can be individually defined by the manufacturer for a particular weighing instrument or for a whole series of instruments as a band defined by a northward and southward latitude, φ_1 and φ_2, and by an upper and lower limit for the height above sea level, h_1 and h_2; see Figure A.7.3.

Figure A.7.3:
Example of a three-dimensional zone according to WELMEC, defined by the latitudes φ_1 and φ_2 as well as the height above sea level h_1 and h_2

For a gravity zone with given values for φ_1, φ_2, h_1 and h_2 the weighing instrument is labelled in the form $\varphi_1 - \varphi_2 \equiv h_1 - h_2$ and adjusted to the average g value (= reference value) according to equation (A.7.1) (but with $a_1 = 9.780318$ according to [89]), whereby the average geographic latitude $\varphi_m = (\varphi_1 + \varphi_2)/2$ and the average height $h_m = (h_1 + h_2)/2$ is used. For the sake of clarity, the manufacturer has to round up the values for φ as multiples of 1° (0.5° is also permitted as an exception) and the values for h as multiples of 100 m. To define a zone or a band, WELMEC has agreed on the criterion $\pm 1/3$ mpe which leads to significantly smaller zones than, for example, in Germany. Therefore, the WELMEC concept is mainly advantageous for countries that currently do not have any gravity zone regulation.

For instruments that are only partially affected by gravitational acceleration such as semi-self-indicating instruments, only the number n' of the scale intervals affected by gravity needs to be considered.

A.8 Gravitational acceleration for selected cities

Table A.8.1 presents gravitational acceleration values g for selected capitals and well-known cities of the world. All of the g values in Table A.8.1 were calculated using the "Gravity Information System" (GIS) provided by the Physikalisch-Technische Bundesanstalt (PTB) [92]. The standard uncertainty of the g values is equal to or less than 0.0002 ms^{-2}.

Continent	Location	Latitude	Longitude	Height above sea level	Gravitational acceleration
		$\varphi\,/\,°$	$\lambda\,/\,°$	$h\,/\,\mathrm{m}$	$g\,/\,\mathrm{ms}^{-2}$
Europe					
	Amsterdam	52.37	4.91	0	9.8127
	Athens	37.98	23.73	90	9.8004
	Barcelona	41.38	2.18	10	9.8031
	Basel	47.57	7.60	260	9.8077
	Belgrade	44.82	20.47	130	9.8058
	Berlin	52.52	13.40	40	9.8128
	Bern	46.95	7.44	540	9.8060
	Bologna	44.50	11.34	50	9.8043
	Bratislava	48.15	17.12	140	9.8087
	Braunschweig	52.27	10.53	70	9.8125
	Bremen	53.08	8.80	10	9.8132
	Brussels	50.84	4.36	70	9.8113
	Budapest	47.50	19.08	100	9.8083
	Cologne	50.93	6.96	50	9.8114
	Copenhagen	55.67	12.58	10	9.8154
	Dresden	51.05	13.75	110	9.8112
	Dublin	53.34	−6.25	20	9.8138
	Duesseldorf	51.22	6.77	40	9.8118
	Erfurt	50.98	11.03	200	9.8111
	Frankfurt / Main	50.11	8.70	100	9.8106
	Geneva	46.20	6.15	380	9.8057
	Genoa	44.42	8.95	20	9.8054
	Hamburg	53.55	10.00	10	9.8137
	Hanover	52.37	9.72	50	9.8126
	Helsinki	60.18	24.93	20	9.8190
	Innsbruck	47.26	11.40	580	9.8054
	Kiel	54.33	10.13	10	9.8145
	Kiev	50.43	30.52	180	9.8105
	Lausanne	46.52	6.63	500	9.8057
	Linz	48.31	14.29	270	9.8082
	Lisbon	38.72	−9.13	10	9.8008
	Ljubljana	46.06	14.51	300	9.8066

London	51.50	−0.12	10	9.8119
Lyon	45.75	4.85	170	9.8063
Madrid	40.39	−3.70	660	9.7997
Magdeburg	52.14	11.63	60	9.8128
Mainz	50.00	8.27	160	9.8105
Milan	45.47	9.19	110	9.8054
Minsk	53.90	27.57	280	9.8136
Moscow	55.75	37.63	150	9.8153
Munich	48.15	11.58	510	9.8072
Naples	40.84	14.25	10	9.8026
Nicosia	35.17	33.37	140	9.7984
Oslo	59.92	10.75	10	9.8192
Palermo	38.12	13.37	10	9.8004
Paris	48.87	2.33	40	9.8094
Potsdam	52.40	13.07	40	9.8127
Prague	50.09	14.44	400	9.8100
Reykjavik	64.15	−21.95	20	9.8225
Rome	41.90	12.48	40	9.8035
Saarbruecken	49.23	7.00	230	9.8094
Salzburg	47.80	13.04	420	9.8068
Schwerin	53.63	11.40	40	9.8137
Sofia	42.68	23.32	550	9.8024
Stockholm	59.33	18.06	20	9.8183
Stuttgart	48.77	9.18	250	9.8087
Turin	45.08	7.68	230	9.8053
Venice	45.44	12.33	0	9.8063
Vienna	48.20	16.37	200	9.8084
Warsaw	52.25	21.00	100	9.8123
Wiesbaden	50.08	8.25	120	9.8105
Zurich	47.38	8.54	410	9.8066

Asia

Ankara	39.93	32.86	900	9.7992
Baghdad	33.34	44.39	40	9.7952
Bangkok	13.75	100.52	10	9.7830
Beijing	39.93	116.39	60	9.8012
Beirut	33.87	35.51	60	9.7967
Calcutta	22.57	88.37	10	9.7879
Chennai (Madras)	13.08	80.28	10	9.7824
Chongqing	29.56	106.55	240	9.7915
Colombo	6.93	79.85	0	9.7812
Damascus	33.50	36.30	690	9.7941
Delhi	28.67	77.22	220	9.7913
Dhaka	23.72	90.41	0	9.7883
Hanoi	21.03	105.85	20	9.7866
Hong Kong	22.28	114.18	30	9.7874

Jakarta	−6.17	106.83	10	9.7815
Jerusalem	31.78	35.22	750	9.7931
Kabul	34.52	69.18	1810	9.7913
Kathmandu	27.72	85.32	1350	9.7870
Kuala Lumpur	3.17	101.70	20	9.7806
Kuwait City	29.37	47.98	10	9.7924
Manila	14.60	120.98	10	9.7834
Mumbai (Bombay)	18.98	72.83	10	9.7863
Riyadh	24.64	46.77	620	9.7878
Seoul	37.57	127.00	50	9.7994
Shanghai	31.22	121.46	10	9.7942
Singapore	1.29	103.86	10	9.7806
Tashkent	41.32	69.25	460	9.8010
Tehran	35.67	51.42	1190	9.7944
Tianjin	39.14	117.18	10	9.8006
Tokyo	35.69	139.75	10	9.7978

Africa

Abuja	9.08	7.53	500	9.7803
Accra	5.55	−0.21	20	9.7809
Addis Ababa	9.02	38.75	2370	9.7746
Algiers	36.76	3.05	10	9.7995
Cairo	30.05	31.25	50	9.7931
Cape Town	−33.92	18.42	10	9.7963
Dakar	14.74	−17.63	10	9.7845
Dar es Salaam	−6.80	39.28	10	9.7809
Freetown	8.49	−13.23	10	9.7821
Harare	−17.82	31.04	1490	9.7812
Johannesburg	−26.20	28.04	1760	9.7853
Kampala	0.32	32.57	1160	9.7770
Khartoum	15.59	32.53	380	9.7829
Lomé	6.13	1.22	60	9.7815
Mbabane	−26.32	31.13	1150	9.7870
Nairobi	−1.28	36.81	1690	9.7752
Niamey	13.52	2.12	200	9.7826
Rabat	34.02	−6.84	20	9.7963
Tripoli	32.89	13.18	10	9.7957
Tunis	36.80	10.18	10	9.7990
Windhoek	−22.57	17.08	1680	9.7830
Yaoundé	3.87	11.52	730	9.7784

North America

Anchorage	61.22	−149.87	30	9.8190
Atlanta	33.75	−84.39	320	9.7951
Boston	42.36	−71.06	40	9.8037

Chicago	41.85	−87.65	180	9.8025
Denver	39.74	−104.98	1600	9.7961
Houston	29.76	−95.36	10	9.7929
Montreal	45.52	−73.65	50	9.8064
Ottawa	45.27	−75.75	70	9.8060
Quebec	46.82	−71.23	10	9.8077
Salt Lake City	40.76	−111.89	1300	9.7978
San Francisco	37.78	−122.40	10	9.7998
Toronto	43.67	−79.38	100	9.8043
Vancouver	49.28	−123.10	10	9.8095
Washington	38.90	−77.04	10	9.8010

Central and South America

Asunción	−25.28	−57.67	50	9.7897
Belo Horizonte	−19.92	−43.93	860	9.7837
Bogota	4.60	−74.08	2620	9.7738
Brasilia	−15.79	−47.90	1150	9.7806
Buenos Aires	−34.59	−58.67	10	9.7969
Caracas	10.50	−66.92	920	9.7804
Guatemala City	14.62	−90.53	1530	9.7795
Havana	23.13	−82.38	10	9.7880
Kingston	18.00	−76.80	50	9.7858
La Paz	−16.50	−68.14	3650	9.7742
Lima	−12.05	−77.05	130	9.7827
Mexico City	19.43	−99.14	2250	9.7794
Montevideo	−34.88	−56.16	30	9.7975
Panama City	8.97	−79.53	10	9.7822
Quito	−0.20	−78.50	2800	9.7726
Rio de Janeiro	−22.90	−43.24	30	9.7880
San José	9.93	−84.08	1160	9.7792
San Salvador	13.70	−89.20	680	9.7818
Santiago de Chile	−33.46	−70.66	540	9.7941
Santo Domingo	18.47	−69.90	10	9.7864
Sao Paulo	−23.53	−46.61	730	9.7864

Australia and Oceania

Adelaide	−34.93	138.60	50	9.7970
Brisbane	−27.48	153.02	40	9.7915
Canberra	−35.30	149.17	600	9.7962
Darwin	−12.45	130.83	10	9.7833
Honolulu	21.31	−157.86	0	9.7895
Melbourne	−37.82	144.97	10	9.7998
Port Moresby	−9.46	147.19	40	9.7819
Sydney	−33.88	151.22	10	9.7966
Wellington	−41.30	174.78	10	9.8027